Quality Control, Robust Design, and the Taguchi Method

The Wadsworth & Brooks/Cole Statistics/Probability Series

Series Editors
O. E. Barndorff-Nielsen, Aarhus University
Peter J. Bickel, University of California, Berkeley
William S. Cleveland, AT&T Bell Laboratories
Richard M. Dudley, Massachusetts Institute of Technology

Quality Control, Robust Design, and the Taguchi Method

Edited by

Khosrow Dehnad
AT&T Bell Laboratories

Wadsworth & Brooks/Cole
Advanced Books & Software
Pacific Grove, California

Printed in the United States of America
10 9 8 7 6 5 4 3 2

Library of Congress Cataloging in Publication Data

Library of Congress Cataloging-in-Publication Data

Quality control, robust design, and the Taguchi method.

    1. Quality control.   2. Engineering design.
3. Experimental design.   I. Dehnad, Khosrow, [date]
TS156.Q3625   1988   658.5′62   88-26220
ISBN 0-534-09048-6

Sponsoring Editor: *John Kimmel*
Editorial Assistant: *Maria Tarantino*
Production Editor: *Nancy Shammas*
Permissions Editor: *Carline Haga*
Cover Design: *Vernon T. Boes*
Cover Printing: *New England Book Components, Inc., Hingham, Massachusetts*
Printing and Binding: *Arcata Graphics/Fairfield, Fairfield, Pennsylvania*

The following institutions and organizations have kindly granted reprint permissions for articles from their publications: American Society for Quality Control, American Statistical Association, American Telephone and Telegraph Company, Cahners Publishing Company, and The Institute of Electrical and Electronics Engineering.

UNIX is a registered trademark of AT&T.
VAX is a trademark of Digital Equipment Corporation.

FOREWORD

In 1980, I received a grant from Aoyama-gakuin university to come to the United States to assist American Industry improve the quality of their products. In a small way this was to repay the help the US had given Japan after the war. In the summer of 1980, I visited the AT&T Bell Laboratories Quality Assurance Center, the organization that founded modern quality control. The result of my first summer at AT&T was an experiment with an orthogonal array design of size 18 (OA18) for optimization of an LSI fabrication process. As a measure of quality, the quantity "signal-to-noise" ratio was to be optimized. Since then, this experimental approach has been named "robust design" and has attracted the attention of both engineers and statisticians.

My colleagues at Bell Laboratories have written several expository articles and a few theoretical papers on robust design from the viewpoint of statistics. Because so many people have asked for copies of these papers, it has been decided to publish them in a book form. This anthology is the result of these efforts.

Despite the fact that quality engineering borrows some technical words from traditional design of experiments, the goals of quality engineering are different from those of statistics. For example, suppose there are two vendors. One vendor supplies products whose quality characteristic has a normal distribution with the mean on target (the desired value) and a certain standard deviation. The other supplies products whose quality characteristic has a uniform distribution with the same mean and standard deviation. The distributions are different from a statistical viewpoint but the products are the same from a quality engineering viewpoint. The reason is that quality engineering measures quality levels by observing the mean square deviation from the nominal or ideal value and relates it to the signal-to-noise ratio.

The philosophy of quality engineering in using experiments with orthogonal array designs is quite different from that of traditional statistical experimental design. Quality engineering uses orthogonal arrays to evaluate new designs for downstream manufacturing. It estimates only the main effects. When only the main effects are estimated using orthogonal arrays, these effects may be confounded by many interactions. When there are no interactions and, hence, when there is no confounding, using other factorial designs is inefficient. Conversely, when many large interactions exist, the design is poor—even useless—because the optimum settings of the design parameters are different downstream.

We should assume that design parameters interact with various conditions: laboratory, downstream, large-scale manufacturing, and customer-use. The most efficient way to conduct experiments for finding the degree of interaction is to conduct them within the laboratory, and then assume the optimum settings of design parameters in the lab is the same for downstream conditions. And to determine whether large interactions exist among product or process design parameters is to compare the predicted output with the outcome of the experiments. When there are large interactions and the predicted value is different from the outcome, we can prevent poor designs from going downstream. When the predicted output and the experiments' outcome agree, we can be confident that the settings are optimum, except for possible adjustments. Orthogonal array experiments, then, can find poor design just as inspections do.

Although I dropped statistical theory as the sole basis for quality engineering years ago, I always welcome any related statistical theory on how to measure and improve quality. I am sure this is the best book to begin the process of advancing quality engineering methodologies and mathematical foundations, including statistical theory. Perhaps the readers of this book will play an important role in this process.

Genichi Taguchi
Tokyo, Japan
December 25, 1986

PREFACE

A decade ago, Dr. Genichi Taguchi was unknown in the United States. Today, he is frequently mentioned with such influential quality leaders as W. Edwards Deming, Joseph M. Juran, and Kaoru Ishikawa. This dramatic change is testimony to the success and the impact of his approach to quality engineering. By combining statistical methods with a deep understanding of engineering problems, Dr. Taguchi has created a powerful tool for quality improvement. The approach focuses on reducing the influence of factors, called *noise*, that degrade the performance of a product or process. This is in contrast to the conventional quality engineering practice of identifying important sources of noise and taking measures, often costly, to control them. Dr. Taguchi's "parameter design" achieves this robustness to noise through proper settings of certain parameters, called *control factors*. Since these parameters are generally easy to set, "macro quality is obtained with micro money." The improved settings of the parameters are obtained through statistically designed experiments that enable one to study a large number of factors with relatively few experiments. Within AT&T, these methods are collectively referred to as "robust design."

The Quality Theory and Technology Department at AT&T Bell Laboratories has worked with Dr. Taguchi on robust design since 1980. Bell Labs researchers have studied and applied this methodology in many cases. The number of requests for reprints of the articles written by these researchers has been growing, as has been the popularity of Dr. Taguchi's approach. Moreover, it has become difficult to follow the evolution of this method because these articles have appeared in a wide variety of publications. This book removes that difficulty by collecting them in a single volume.

The first test of parameter design at Bell Labs was in the summer of 1980. Earlier that year, Dr. Taguchi had expressed an interest in visiting Bell Labs to discuss his ideas on both quality engineering and the design of experiments. An invitation was extended to him, and he visited the Quality Assurance Center at Bell Labs during August, September, and October of 1980.

Some of the ideas and approaches proposed by Dr. Taguchi were novel and, at times, controversial and difficult to understand; the language barrier undoubtedly contributed to the difficulty. He defined quality as "the monetary loss to the society from the time a product is shipped." This ran counter to the view which commonly associates quality with something good and desirable. Dr. Taguchi also advocated making a system robust to factors (noise) that degraded the system's performance. This idea was contrary to the practice of finding and eliminating assignable causes of poor quality. He used statistically planned experiments to study the effect of different factors on the robustness of the system. And he measured these effects with statistics he called signal-to-noise ratios—to this day, a controversial issue. This use was somewhat novel because designed experiments have been mainly used to determine the impact of factors on the system's average performance—the "main effects."

It soon became evident that the merits of his ideas and approaches could not be judged solely on theoretical grounds; thus, Dr. Taguchi volunteered to test his approach in an actual case. The case selected was the photolithography process used to form contact windows on silicon wafers. In this process, which is similar to photography, photoresist is applied to a wafer and dried by baking. The wafer is exposed to ultraviolet through a mask, and the photoresist is removed from the exposed areas, which are the future windows. These areas are etched in a high-vacuum chamber, and the remaining photoresist is removed.

Using Dr. Taguchi's approach, in eight weeks and with only 18 experiments, nine important process parameters were studied. The analysis of the results revealed improved settings of the parameters, and the process was adjusted accordingly. The test case was a resounding success. The improvements were excellent when measured against the resources and time spent on the project. Specifically, the variance of pre-etch line width, routinely used as a process quality indicator, was reduced by a factor of four. In addition, the defect density due to unopened or unprinted windows per chip was reduced by a factor of three. Finally, the overall time spent by wafers in window photolithography was reduced by a factor of two. This last improvement was because the process engineers, by observing these improvements over several weeks, gained confidence in the stability and robustness of the new process and were able to eliminate a number of in-process checks. (For details, see the first article in the **Case Studies** section.)

The remarkable success of the test case encouraged further application of Dr. Taguchi's techniques and initiated research at Bell Labs designed to elucidate, modify, and enhance his approach. This research effort was joined in 1985 by the University of Wisconsin's Center for Quality and Productivity, under the directorship of Professor George E. P. Box. The opportunity for this collaboration between academia and industry was created by a research grant from the National Science Foundation. The results of the research at Bell Labs and the Center for Quality and Productivity have been presented at conferences and seminars both in the United States and abroad, and have appeared in journals, conference proceedings, and technical reports. The articles by Bell Labs authors that have appeared cover various aspects of Dr. Taguchi's approach. Some of the articles are expository in nature, some are case studies, and some are discussions of technical issues.

The present anthology shows the evolution to date of robust design at Bell Labs and illustrates the fact that the field is still evolving, with controversial issues and differences of opinion among researchers and practitioners. In addition to these facts, the book provides an overview of the field for people, such as managers, who desire a general understanding of the subject and of its technical potentials and benefits. They can use this knowledge to lead the way in the application of robust design techniques in their own organizations. This book also serves as a guide for practitioners, such as product or process engineers, who might want to apply robust design techniques. Finally, it presents the results of research in this field and some of the open issues deserving further study.

To cover the full breadth of the areas covered by the articles and to serve the needs of audiences named above, the book contains three sections: **Overview**, **Case Studies**, and **Methodology**.

The **Overview** section provides a general understanding of the methodology. It starts with Article 1, which describes and interprets Dr. Taguchi's quality philosophy. This first article covers his definition of quality in terms of societal costs and their relation to variability. It also describes the use of quadratic functions to approximate these costs, the need for continuous reduction in variability, and the use of statistically planned experiments as a way of achieving this reduction. Article 2 is an exposition of the main ideas of robust design through a case study. With the help of examples, Article 3 describes the use of statistically planned experiments in quality engineering and their potential as a tool for quality improvement. Article 4 is a comprehensive description of what has come to be known as the Taguchi Method. The name of the book is partly borrowed from the title of this article, "Off-Line Quality Control, Parameter Design, and the Taguchi Method." The final article of this section, Article 5, reviews the elements of parameter design. It divides engineering problems into general classes and suggests, without proofs, a signal-to-noise ratio for each class.

The second section, **Case Studies**, is intended for the practitioner. It starts with Article 6, which gives details of the first case study at Bell Labs. It also illustrates "accumulation analysis," the method Dr. Taguchi recommends for analyzing data that are ordered categories, such as poor, average, and good. Article 7 discusses the application of parameter design to improve the fluxing-soldering-cleaning of the wave soldering process. It also serves as a good introduction to wave soldering. Next, Article 8 reviews parameter design and presents a four-step operational procedure for its application. The methodology and the steps are illustrated using the example of growing an epitaxial layer on silicon wafers. Article 9 demonstrates the application of statistically planned experiments to improve the response time of a computer. And Article 10, the final article of the **Case Studies** section, discusses two parameter design case studies: router bit life improvement, and optimizing a differential operational amplifier circuit.

The final section, **Methodology**, presents some of the research on robust design conducted at Bell Labs. It starts with Article 11 on accumulation analysis. This article discusses the properties of Taguchi's accumulation analysis and points out that the degrees of freedom that Dr. Taguchi uses are approximate. It also shows that the method does reasonably well in detecting location effects but is unnecessarily complicated. The author then proposes two simple scoring methods for detecting dispersion and location effects separately. Next, Article 12 gives a mathematical formulation of parameter design problems and defines the notion of

"Performance Measures Independent of Adjustment" (PERMIA). The authors show that, under certain assumptions, Dr. Taguchi's signal-to-noise ratios are PERMIAs. They point out that, in the absence of certain assumptions, the use of the SN ratios proposed in Article 5 does not necessarily lead to minimization of the expected loss. Article 13 demonstrates that, for certain models and under a quadratic loss assumption, Dr. Taguchi's two-step optimization has a natural geometric interpretation. Finally, Article 14, the last article of the book, is a road map for analyzing data from quality engineering experiments. The strategy offered has three phases: exploration, modeling, and optimization.

Today, quality has emerged as an important strategic weapon in the marketplace. American industry is generally heeding this message and is responding to the challenge of providing quality products and services at competitive prices. But to meet this challenge even more effectively, industry must take advantage of every technique to improve the quality of its products. The application of robust design provides an ideal opportunity. Until now, to read about this methodology meant sifting through many different publications. This anthology helps remove that obstacle and makes the methodology more accessible. It is hoped that this book will encourage further research in the field of robust design.

Acknowledgments

This book would not be possible without the help of many people and organizations. I would like to thank the authors for their help, suggestions, and contributions. The support of the Quality Assurance Center of AT&T Bell Laboratories and its director Edward Fuchs is gratefully acknowledged. Special thanks is given to Jeffrey Hooper and Blanton Godfrey, the present and former heads of the Bell Laboratories Quality Theory and Technology Department, for their constant help and support. Sigmund Amster, of the Quality Theory and Technology Department, helped in resolving some particularly difficult issues; his help is greatly appreciated. The typesetting of the book was done by the Publication Center of AT&T Bell Laboratories. Mari-Lynn Hankinson oversaw the initial phase of the production, and Robert Wright, Kathleen Attwooll, and Marilyn Tomaino were responsible for the latter part of it. Their excellent help and the help of others at the Publication Center is greatly appreciated. Thanks also to John Kimmel from Wadsworth & Brooks/Cole for his help and patience.

Editorial Method

The articles in this book appear in their original form, with only minor editorial changes, and are self-contained so that they can be read selectively. To help the reader select articles, the table of contents includes the abstract of each article.

An important criterion for the arrangement of the articles has been to make the book suitable for a reader who has no prior knowledge of robust design and the Taguchi Method. Such a reader, by reading the articles consecutively, is led from introductory topics to more advanced and specialized ones. However, because the articles are self-contained, their introductory sections sometimes cover ground that might have been covered before. In such cases, the reader can review those introductory sections quickly.

If an article refers to a source that appears in the book, that source is indicated by cross reference in the article's references.

The articles have been written by different authors at different times on a subject that is still evolving. Consequently, there are some variations among the terms, a source of possible confusion at times. To remedy this problem without compromising the spirit of the anthology, a glossary of terms is included at the beginning of the book. Should an author use a term different from that defined in the glossary, the first occurrence of that term is indicated in a footnote together with its glossary equivalent.

Khosrow Dehnad

CONTENTS

PART ONE - OVERVIEW   1

1 TAGUCHI'S QUALITY PHILOSOPHY: ANALYSIS AND COMMENTARY   3

Raghu N. Kackar

The popularity of Genichi Taguchi testifies to the merit of his quality philosophy. However, a lack of proper communication has kept some of his ideas in a shroud of mystery. This chapter describes the following elements of Taguchi's quality philosophy: society's view of quality; the importance of quality improvement; the need for reducing performance variation; customer's loss due to performance variation; importance of product and process designs; off-line quality control; and parameter design experiments.

2 MACRO-QUALITY WITH MICRO-MONEY   23

Lewis E. Katz
Madhav S. Phadke

A new quality engineering technique called the "robust design method" drastically reduces the number of time-consuming tests needed to determine the most cost-effective way to manufacture microchips while maintaining high quality standards. Robust design involves the use of mathematics to improve the cost-effectiveness of manufacturing processes. Cost-effectiveness includes both time and money, and both must be kept to an absolute minimum. Robust design makes the production process insensitive to variation so that there are few surprises when the new chip is placed into production. The backbone of the robust design concept is the orthogonal array method of simultaneously studying many process variables. A suitable orthogonal array is used to determine how many tests should be conducted and with which values of process variables. At the end of the study the effect of each variable can be separated from the others.

3 QUALITY ENGINEERING USING DESIGN OF EXPERIMENTS   31

Madhav S. Phadke

By quality engineering we mean reducing the deviation of a product's functional characteristic from the target value. The cost-effective way to minimize these deviations is to design the product and its manufacturing process so that the product performance is least sensitive to the noise, internal and external to the product. This problem involves selecting levels of a large number of variables through experimentation of simulation. Professor Genichi Taguchi has developed some simple and elegant methods of constructing fractional factorial designs using orthogonal arrays and linear graphs. He has also developed some special methods of analyzing the resulting data. In this chapter we will present these methods, which are widely used by engineers in Japan, and some applications in AT&T Bell Laboratories.

4 OFF-LINE QUALITY CONTROL, PARAMETER DESIGN, AND THE TAGUCHI METHOD   51

Raghu N. Kackar

Off-line quality control methods are quality and cost control activities conducted at the product and process design stages to improve product manufacturability and reliability, and to reduce product development and lifetime costs. Parameter design is an off-line quality control method. At the product design stage the goal of parameter design is to identify settings of product design characteristics that make the product's performance less sensitive to the effects of environmental variables, deterioration, and manufacturing variations. Because parameter design reduces performance variation by reducing the influence of the sources of variation rather than by controlling them, it is a very cost-effective technique for improving product quality. This chapter introduces the concepts of off-line quality control and parameter design and then discusses the Taguchi Method for conducting parameter design experiments.

Genichi Taguchi
Madhav S. Phadke

The main technological reason for the success of Japan in producing high–quality products at low cost is the emphasis placed on the optimization of the product design and process design for manufacturability, quality and reliability, and the availability of efficient experimentation and simulation methods for performing optimization. The role of design optimization is to minimize sensitivity to all noise factors—external noise, manufacturing variation, and deterioration of parts. It involves: 1) selecting an appropriate objective function; 2) evaluating the objective function; and 3) maximizing the objective function with respect to a number of decision variables called control factors. This chapter describes the general design optimization problem and shows how to select the objective function in a variety of engineering problems. Also, through a circuit design example, we show how orthogonal arrays can be used to systematically and efficiently evaluate and maximize the objective function.

Madhav S. Phadke   D. V. Speeney
Raghu N. Kackar   M. J. Grieco

In this chapter we describe the off-line quality control method and its application in optimizing the process for forming contact windows in 3.5-μm complementary metal-oxide semiconductor circuits. The off-line quality control method is a systematic method of optimizing production processes and product designs. It is widely used in Japan to produce high-quality products at low cost. The key steps of off-line quality control are: 1) identify important process factors that can be manipulated and their potential working levels; 2) perform fractional factorial experiments on the process using orthogonal array designs; 3) analyze the resulting data to determine the optimum operating levels of the factors (both the process mean and the process variance are considered in this analysis; and 4) conduct an additional experiment to verify that the new factor levels indeed improve the quality control.

7   OPTIMIZING THE WAVE SOLDERING PROCESS   143

K. M. Lin
Raghu N. Kackar

Wave soldering of circuit pack assemblies involves three main phases: fluxing, solder-ing, and cleaning. The function of each of these three phases is, in turn, dependent upon a number of factors, and variations in any of these factors can drastically affect the result of soldering process and circuit pack assembly reliability. The orthogonal array design method was employed in setting up a highly fractionated, designed exper-iment to optimize the entire fluxing-soldering-cleaning process. In the initial experi-ment, a total of 17 factors were studied in only 36 test runs. Results from this experi-ment were then used to set up some smaller scale follow-up experiments to arrive at a robust wave-soldering process which includes an excellent, new flux formulation as well as the proper soldering and cleaning procedures.

8   ROBUST DESIGN: A COST-EFFECTIVE METHOD FOR IMPROVING MANUFACTURING PROCESSES   159

Raghu N. Kackar
Anne C. Shoemaker

Robust design is a method for making a manufacturing process less sensitive to manufacturing variations. Because it reduces variation by reducing the influence of sources of variation, not by controlling them, robust design is a cost-effective technique for improving process quality. The method uses small, statistically planned experi-ments to vary the settings of key process control parameters. For each combination of control parameter settings in the experiment, process performance characteristics are measured to reflect the effects of manufacturing variation. Then, simple plots or tables are used to predict the control parameter settings that minimize these effects. A small follow-up experiment confirms the prediction. This chapter describes how to apply the basic robust design method to improve process quality.

9   TUNING COMPUTER SYSTEMS FOR MAXIMUM PERFORMANCE: A STATISTICAL APPROACH   175

William A. Nazaret
William Klingler

In this chapter we discuss a statistical approach to the problem of setting the tunable parameters of an operating system to minimize response time for interactive tasks. This approach is applicable to both tuning and benchmarking of computer systems. It is based on the use of statistical experimental design techniques and represents an inexpensive, systematic alternative to traditional ways of improving the response of systems. We illustrate the method by means of experiments performed on VAX minicomputers running under the UNIX System V operating system.

10 DESIGN OPTIMIZATION CASE STUDIES   187

Madhav S. Phadke

Designing high-quality products at low cost is an economic and technological challenge to the engineer. A systematic and efficient way to meet this challenge is a new method of design optimization for performance, quality, and cost. The method, called robust design, has been found effective in many areas of engineering design. In this chapter, the basic concepts of robust design will be discussed and two applications will be described in detail. The first application illustrates how, with a very small number of experiments, highly valuable information can be obtained about a large number of variables for improving the life of router bits used for cutting printed wiring boards from panels. The second application shows the optimization of a differential operational amplifier circuit to minimize the dc offset voltage by moving the center point of the design, which does not add to the cost of making the circuit.

PART THREE - METHODOLOGY   213

11 TESTING IN INDUSTRIAL EXPERIMENTS WITH ORDERED CATEGORICAL DATA   215

Vijayan N. Nair

This chapter deals with techniques for analyzing ordered categorical data from industrial experiments for quality improvement. Taguchi's accumulation analysis method is shown to have reasonable power for detecting the important location effects; however, it is an unnecessarily complicated procedure. For detecting dispersion effects, it need not even be as powerful as Pearson's chi-square test. Instead, two simple and easy-to-use scoring schemes are suggested for separately identifying the location and dispersion effects. The techniques are illustrated on data from an experiment to optimize the process of forming contact windows in complementary metal-oxide semiconductor circuits.

Ramon V. Leon
Anne C. Shoemaker
Raghu N. Kackar

The connection between quadratic loss and Taguchi's signal-to-noise ratios leads to a general principle for choosing performance measures in parameter design. Taguchi states that the objective of parameter design is to find the settings of product or process design parameters that minimize average quadratic loss, i.e., the average squared deviation of the response from its target value. Yet in practice to choose the settings of design parameters he maximizes a set of measures called signal-to-noise ratios. In general, he gives no connection between these two optimization problems. In this chapter we show that *if certain models for the product or process response are assumed, then the maximization of the signal-to-noise ratio leads to the minimization of average squared error loss.* The signal-to-noise ratios take advantage of the existence of special design parameters, called adjustment parameters. When these parameters exist, use of the signal-to-noise ratio allows the parameter design optimization procedure to be conveniently decomposed into two smaller optimization steps, the first being maximization of the signal-to-noise ratio. However, we show that under different models (or loss functions) other performance measures give convenient two-step procedures, while the signal-to-noise ratios do not. We call performance measures that lead to this convenient two-step procedures Performance Measures Independent of Adjustment. In this language Taguchi's signal-to-noise ratios are special cases of these performance measures.

Khosrow Dehnad

The notion of quadratic loss is used to arrive at a formal definition of Taguchi's signal-to-noise ratio. This chapter shows that this ratio can be interpreted as a measure of the angular deviation of a product or process functional characteristic from the desired value of this characteristic. A geometric interpretation of the decomposition of loss into adjustable and nonadjustable components are presented along with the two-step optimization procedure on which it is based.

14 A DATA ANALYSIS STRATEGY FOR
QUALITY ENGINEERING EXPERIMENTS 289

Vijayan Nair
Daryl Pregibon

This chapter deals with techniques for analyzing data from quality engineering experiments in order to optimize a process with fixed target. We propose a structured data-analytic approach with three phases of analysis: an exploratory phase, a modeling phase, and an optimization phase. We emphasize the use of graphical methods in all three phases to guide the analysis and facilitate interpretation of results. We discuss the role of data transformations and the relationship between an analysis of transformed data and Taguchi's signal-to-noise ratio analysis. Our strategy is presented in a form that can be easily followed by users and the various steps are illustrated by applying them to an example.

GLOSSARY OF TERMS USED IN ROBUST DESIGN

R. V. León
A. C. Shoemaker

This glossary contains a set of standard definitions sponsored by AT&T Bell Laboratories Quality Assurance Center. It is hoped that these definitions will facilitate communication among practitioners and users of parameter design. It represents the efforts of many members of Bell Laboratories. Thanks are due to K. Dehnad, J. W. Schofield, P. G. Sherry and G. Ulrich. Special thanks are due M. S. Phadke who provided particularly valuable and insightful comments. For a discussion of parameter design, see Kackar.[1]

DEFINITIONS

Adjustment parameter. A control parameter which can be used to fine-tune performance after the settings of nonadjustment control parameters have been chosen to minimize a performance measure independent of adjustment.

Control array. An array whose rows specify the settings of the control parameters in an experiment. Taguchi calls this array an "inner array".

Control parameter. A variable(s) whose nominal settings can be controlled by the designers or process engineers.

Functional characteristic. The basic characteristic(s) of a product that is related to its functionality.

Mean adjustment parameter. An adjustment parameter(s) which controls the average value of the functional characteristic.

Noise (in parameter design). Variations that cause a product's performance characteristics to fluctuate when the nominal values of control parameters are fixed. Examples include product deterioration, imperfections in raw materials, and variation in the manufacturing environment or the customer's environment.

Noise array. An array whose rows specify the settings of noise factors in an experiment. Taguchi calls this array an "outer array".

Noise parameter. A source(s) of noise that can be systematically varied in a parameter design experiment.

Off-line quality control method. The method(s) used during product or manufacturing process design to guarantee or improve the quality of a product.

On-line quality control method. The method(s) used during manufacturing to improve and maintain the quality of a product.

Orthogonal array. An array of numbers whose columns are pairwise orthogonal; that is, in every pair of columns, all ordered pairs of numbers occur an equal number of times.

Performance loss. Loss caused by deviation of a functional characteristic from its target value.

Performance measure. A measure of average performance loss caused by the deviation of a functional characteristic from its target value. Because it is an average over many units of a product, a performance measure quantifies the quality of the *design* of a product or process, rather than the quality of an individual unit. The objective of parameter design is to find settings of control parameters that optimize the performance measure.

Performance measure independent of adjustment (PERMIA). A performance measure which is independent of the settings of adjustment parameters. Use of a PERMIA allows parameter design to be done in two steps: First identify the settings of nonadjustment control parameters to minimize the PERMIA. Then identify the settings of the adjustment control parameters that minimize the performance measure, while nonadjustment design parameters are set at the values found in the first step.

Performance statistic. A statistical estimator of a performance measure.

Robust design. An approach to designing a product or process that emphasizes reduction of performance variation through the use of design techniques that reduce sensitivity to sources of variation.

Robustness measure. A performance measure which quantifies the variation of a product's functional characteristic about its average value.

Robustness statistic. A statistical estimator of a robustness measure.

Scale adjustment parameter. An adjustment parameter(s) which controls the magnitude of the linear effect of a signal parameter on the functional characteristic.

Signal parameter. A variable which is used to change the value of the functional characteristic to attain a desired value. The designer does not choose the setting of this parameter; however, he or she may wish to design the product or process to be very sensitive to changes in the signal parameter.

Signal to noise ratios (in parameter design). A set of performance statistics proposed by Genichi Taguchi (see Taguchi and Wu[2]).

Stages of product or manufacturing process design (quality engineering view).

1. **System design**. The process of applying scientific and engineering knowledge to produce a basic working prototype design.

2. **Parameter design**. The process of identifying product or manufacturing process control parameter settings that reduce sensitivity to sources of variation.

3. **Tolerance design**. The process of selecting economically justifiable tolerances for the product or process control parameters.

Target value. The desired value of a performance characteristic.

REFERENCES

1. Kackar, Raghu N. 1985. "Off-Line Quality Control, Parameter Design, and the Taguchi Method." *Journal of Quality Technology* **17**(Oct.):176-209.
                                    *(Quality Control, Robust Design, and
                                    the Taguchi Method; Article 4)*

2. Taguchi, Genichi, and Yu-In Wu. 1980. *Introduction to Off-Line Quality Control.* Tokyo: Central Japan Quality Control Association. Available from the American Supplier Institute, 6 Parklane Blvd., Suite 411, Dearborn, MI 48126

ADDITIONAL REFERENCES

Kackar, Raghu N., and Anne C. Shoemaker. 1986. "Robust Design: A Cost-Effective Method for Improving Manufacturing Processes." *AT&T Technical Journal* (April).

(Quality Control, Robust Design, and the Taguchi Method; Article 8)

Leon, Ramon V., Anne C. Shoemaker, and Raghu N. Kackar. 1987. "Performance Measures Independent of Adjustment: An Explanation and Extension of Taguchi's Signal-to-Noise Ratios." *Technometrics* vol. 29, no. 3, (Aug.):253-265.

(Quality Control, Robust Design, and the Taguchi Method; Article 12)

Taguchi, Genichi, and Madhav S. Phadke. 1984. "Quality Engineering Through Design Optimization." *Conference Record, GLOBECOM84 Meeting.* Atlanta, Georgia: IEEE Communications Society. 1106-1113.

(Quality Control, Robust Design, and the Taguchi Method; Article 5)

PART ONE

OVERVIEW

1

TAGUCHI'S QUALITY PHILOSOPHY:
ANALYSIS AND COMMENTARY

An Introduction to and Interpretation of Taguchi's Ideas

Raghu N. Kackar

1.1 INTRODUCTION

Today, Genichi Taguchi is frequently mentioned along with W. Edwards Deming, Kaoru Ishikawa, and J. M. Juran. His popularity testifies to the merit of his quality philosophy. However, a lack of proper communication has kept some of his ideas in a shroud of mystery. This paper will introduce the readers to some of the basic elements of Taguchi's quality philosophy.

Seven points explain the basic elements of Genichi Taguchi's quality philosophy in a nutshell.

1. An important dimension of the quality of a manufactured product is the total loss generated by that product to society.

2. In a competitive economy, continuous quality improvement and cost reduction are necessary for staying in business.

3. A continuous quality improvement program includes incessant reduction in the variation of product performance characteristic about their target values.

4. The customer's loss due to a product's performance variation is often approximately proportional to the square of the deviation of the performance characteristic from its target value.

5. The final quality and cost of a manufactured product are determined to a large extent by the engineering designs of the product and its manufacturing process.

6. A product's (or process') performance variation can be reduced by exploiting the nonlinear effects of the product (or process) parameters on the performance characteristics.

7. Statistically planned experiments can be used to identify the settings of product (and process) parameters that reduce performance variation.

These seven points do not cover all of Taguchi's ideas. Some of these points have also been made by other quality experts.

The following sections discuss each of these points in detail. Most of this discussion represents my interpretations and extensions of Taguchi's ideas. I have also made an attempt to relate these ideas to the published literature.

1.2 SOCIETAL VIEW OF QUALITY

An important dimension of the quality of a manufactured product is the total loss generated by that product to society. Quality is a complex and multifaceted concept. Garvin has surveyed eight different aspects of quality: performance, features, reliability, conformance, durability, serviceability, aesthetics, and perceived quality.[1] The importance of a particular aspect of quality changes with the nature of the product and the needs of the customer. Therefore the specific meaning of the word quality changes with the context in which it is being used. Perhaps no single definition of quality can encompass all possible ideas this word can convey.

Taguchi has drawn our attention to a very important dimension of quality: the societal loss caused by the product.[2] According to Taguchi, "Quality is the loss imparted to the society from the time a product is shipped." This is a strange definition because the word quality connotes desirability, while the word loss conveys the idea of undesirability. The essence of his words is that the societal loss generated by a product, from the time a product is shipped to the customer, determines its desirability. The smaller the loss, the more desirable is the product. Examples of societal losses from a product include: failure to meet the customer's requirements of fitness for use, failure to meet the ideal performance, and harmful side effects caused by the product. All societal losses due to poor performance of a product should be attributed to the quality of the product.

Taguchi's definition seems incomplete. How about the losses to the society while the product is being manufactured? The raw materials, energy, and labor consumed in producing unusable products are societal losses. Also, the toxic chemicals produced during manufacturing can cause losses to the society.

Taguchi's definition should be extended to include the societal losses during manufacturing.

The societal view of quality is a profound concept. According to this concept, the aim of quality control is to reduce the total societal cost, and the function of quality control is to discover and implement innovative techniques that produce net savings to society. An effective quality control program saves society more than it costs and it benefits everybody.

The concept of a societal loss gives us a new way to think about investments in quality improvement projects. Investment in a quality improvement project is justified as long as the resulting savings to customers are more than the cost of improvements. Let us consider an example. If a manufacturer ships a product that imparts a loss of $1,000 to a customer, while the cost to the manufacturer of preventing the customer's loss is only $200, the whole society loses $800 ($1,000-$200). The prevention cost of $200 not spent by the manufacturer can cost him customer's confidence and good will. And this cost can result in a loss of market share worth many millions of dollars. Investments in quality improvement projects appear much more attractive when one takes a long-term view.

1.3 IMPORTANCE OF QUALITY IMPROVEMENT

In a competitive economy, continuous quality improvement and cost reduction are necessary for staying in business. In a competitive economy, a business that does not earn a reasonable profit cannot survive for long. Profit is the difference between the selling price and the manufacturing (and marketing) cost per unit times the number of units sold. When customers are well informed and have a free choice, the selling price is primarily determined by the selling price of the competitor's comparable products and by other market conditions. Therefore profit is determined by the number of units sold (or market share) and the manufacturing (and marketing) cost.

A sure way of increasing the market share is to provide high quality products at low price. Customers want both high quality and low price. Nevertheless, most customers don't mind paying a little bit more for a higher quality product. Indeed quality has no meaning without reference to price. For example, it is unfair to compare the quality of a Chevrolet with that of a Mercedes-Benz because of their substantial price difference. But who would buy a Chevrolet if a Mercedes-Benz were available for the same price?

Companies determined to stay in business use high quality and low cost as their competitive strategy. Such companies know that quality is never good enough and manufacturing cost is never low enough, and they incessantly improve quality and reduce manufacturing cost of their products. Also, in the new global economy, well-informed customers have ever-rising expectations. Therefore, in a free enterprise system, it is necessary to improve quality and reduce manufacturing cost on a continuous basis.

1.4 NEED FOR REDUCING PERFORMANCE VARIATION

A continuous quality improvement program includes incessant reduction in the variation of product performance characteristics about their target values. A product's quality cannot be improved unless the product's quality characteristics can be identified and measured. Further, a continuous quality improvement program depends on knowledge of the ideal values of these quality characteristics. Each quality characteristic is a variable entity. Its value can be different for different units of the product, and also it can change over time for a given unit. The object of a continuous quality improvement program is to reduce the variation of the product's quality characteristics about their desired values.

Almost all products have numerous quality characteristics. But it is neither economical nor necessary to improve all quality characteristics because not all quality characteristics are equally important. It is sufficient to improve the primary quality characteristics. *Performance characteristics* of a product are the primary quality characteristics that determine the product's performance in satisfying the customer's requirements. The sharpness of the picture on a TV set is an example of a performance characteristic. The ideal value of a performance characteristic is called the *target value*. A high quality product performs near the target value consistently throughout the product's life span and under all different operating conditions. For example, a TV set whose picture quality varies with weather conditions has poor quality. Likewise an automobile tire with an extended life span of 40,000 miles has poor quality if it wears out in only 20,000 miles. The variation of a performance characteristic about its target value is referred to as *performance variation*. The smaller the performance variation about the target value, the better is the quality.

Performance variation can be evaluated most effectively when the performance characteristic is measured on a continuous scale. This is because continuous measurements can detect small changes in quality. Therefore it is important to identify performance characteristics of a product that can be measured on a continuous scale. Some performance characteristics of an automobile that can be measured on a continuous scale are: the amount of carbon monoxide produced by the exhaust, the braking distance to make a complete stop for a given speed and brake pedal pressure, the time to accelerate from zero to 55 mph, and the amount of noise caused by the engine.

All target specifications of continuous performance characteristics should be stated in terms of nominal levels and tolerances about the nominal levels. It is a widespread practice in the industry to state target values in terms of interval specifications only. This practice conveys an erroneous idea: that the quality level remains the same for all values of the performance characteristic in the specification interval and then suddenly deteriorates the moment performance value slips out of the specification interval. The target value should be defined as the ideal state of the performance characteristic. Note that the target value need not be the midpoint of the tolerance interval.

Some performance characteristics cannot be measured on a continuous scale, either because of the nature of the performance characteristic or because of limitations of the measuring methods. For example, performance characteristics that require subjective evaluation cannot be measured on a continuous scale. The next best thing to continuous measurements is measurements on an ordered categorical scale such as: poor, fair, good, and excellent. Ordered categorical measurements approximate continuous measurements in the same way that a histogram approximates a continuous distribution. When a measuring instrument has finite range and some of the values are out-of-range, the data are a mixture of discrete and continuous values. Such data can be approximated by an ordered categorical distribution. Similarly, the censored data from reliability tests can also be approximated by an ordered categorical distribution.

Although ordered categorical measurements are less effective in detecting small changes in quality than continuous measurements, they are more effective than binary measurements such as good or bad. For example, classification of solder bonds as good or bad provides meager information about the quality of solder bonds. A more informative classification would be based on the amount of solder in the joint; no solder, insufficient solder, good solder, and excess solder.

1.5 CUSTOMER'S LOSS DUE TO PERFORMANCE VARIATION

The customer's loss due to a product's performance variation is often approximately proportional to the square of the deviation of the performance characteristic from its target value. Any variation in a product's performance characteristic about its target value causes a loss to the customer. This loss can range from mere inconvenience to monetary loss and physical harm. Let Y be a performance characteristic measured on a continuous scale and suppose the target value of Y is τ. Let $l(y)$ represent losses in terms of dollars suffered by an arbitrary customer at an arbitrary time during the product's life span due to the deviation of Y from τ. Generally, the larger the deviation of the performance characteristic Y from its target value τ, the larger is the customer's loss $l(y)$. However, it is usually difficult to determine the actual form of $l(y)$. Often, a quadratic approximation to $l(y)$ adequately represents economic losses due to the deviation of Y from τ. (The use of quadratic approximations is not new. Indeed, quadratic approximation is the basis of the statistical theory of least squares developed by Gauss in 1809.) The simplest quadratic loss function is

$$l(Y) = k(Y - \tau)^2,$$

where k is some constant. (See Figure 1.1.) Then unknown constant k can be determined when $l(Y)$ is known for a particular value of Y. For example, suppose that

$$(\tau-\Delta, \tau+\Delta)$$

is the customer's tolerance interval, that a product performs unsatisfactorily when Y slips out of this interval, and that the cost to the customer of repairing or discarding the product is A dollars. Then

$$A = k\Delta^2 \text{ and } k = A/\Delta^2.$$

This version of the loss function is useful when a specific target value is the best and the loss increases symmetrically as the performance characteristic deviates from the target. But the concept can be extended to many other situations. Two special cases will be discussed later: the smaller the better (this would be the case, for example, when the performance characteristic is the amount of impurity and the target value is zero; the smaller the impurity, the better it is), and the larger the better (this would be the case, for example, when the performance characteristic is the strength; the larger the strength, the better it is).

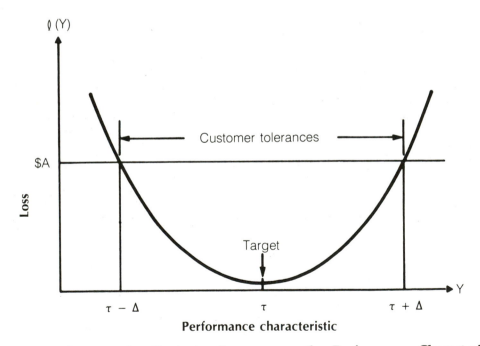

Figure 1.1. **Loss to the Customer Increases as the Performance Characteristic Deviates from the Target Value**

The average loss to customers due to performance variation is obtained by "statistically averaging" the quadratic loss

$$l(Y) = k(Y - \tau)^2$$

associated with the possible values of Y. In the case of quadratic loss functions, the average loss due to performance variation is proportional to the *mean squared error* of Y about its target value τ. Therefore the fundamental measure of variability is the mean squared error and not the variance.

The concept of average loss due to performance variation can also be used to characterize process capability independently of the process specification limits. Process capability is often characterized by the "percentage nonconforming" which depends on the process specification limits. But process specification limits are only tentative cut-off points used for standardizing a manufacturing process. As the process variability is reduced, the specification limits are often tightened. The average loss due to performance variation provides a measure of process variation that is independent of the tentative specification limits. The nonconformance rates can be reduced to ppm (parts per million) levels only when the specification limits are interpreted as tentative cut-off points and the variation in underlying performance characteristics is reduced continuously. The concept of quadratic loss emphasizes the importance of continuously reducing performance variation.

1.6 IMPORTANCE OF PRODUCT AND PROCESS DESIGNS

The final quality and cost of a manufactured product are determined to a large extent by the engineering designs of the product and its manufacturing process. A product's development cycle can be divided into three overlapping stages: product design, process design, and manufacturing. Each product development stage has many steps. The output of one step is the input to the next. Therefore all steps, especially the transfer points, affect the final quality and cost. However, because of the increasing complexity of modern products, product and process designs play crucial roles. Indeed the dominance of a few corporations in high technology products such as automobiles, industrial robots, microprocessors, optical devices, and machine tools can be traced to the strengths of the engineering designs of their products and manufacturing processes.

The importance of product design was highlighted by a mishap that occurred a few years ago when an important U.S. military mission, to rescue the hostages in Iran, failed because the helicopters got bogged down in a dust storm. The dust destroyed the bearings and clogged the engines of the helicopters. This catastrophe would not have occurred if the helicopters had been designed to withstand severe dust storms. Generally, a product's field performance is affected by environmental variables (and human variations in operating the product), product deterioration, and manufacturing imperfections. (Note that these sources of

9

variation are chronic problems.) Manufacturing imperfections are the deviations of the actual parameters of a manufactured product from their nominal values. Manufacturing imperfections are caused by inevitable uncertainties in a manufacturing process and they are responsible for performance variation across different units of a product. As indicated in Table 1.1, countermeasures against performance variation caused by environmental variables and product deterioration can be built into the product only at the product design stage.

TABLE 1.1. PRODUCT DEVELOPMENT STAGES AT WHICH COUNTER-MEASURES AGAINST VARIOUS SOURCES OF VARIATION CAN BE BUILT INTO THE PRODUCT

Product Development Stages	Sources of Variation		
	Environmental Variables	Product Deterioration	Manufacturing Variations
Product design	0	0	0
Process design	X	X	0
Manufacturing	X	X	0

0–Countermeasures possible
X–Countermeasures impossible

The manufacturing cost and manufacturing imperfections in a product are largely determined by the design of the manufacturing process. With a given process design, increased process controls can reduce manufacturing imperfections. (See Figure 1.2.) But process controls cost money. (Nevertheless the cost of process controls is justified as long as the loss due to manufacturing imperfections is more than the cost of process controls). Therefore it is necessary to reduce both manufacturing imperfections and the need for process controls. And this can be accomplished only by improving the process design to move the cost curve down.

The importance of process design is not an entirely new idea. Shewhart[3] must have realized the importance of process design. He emphasized the importance of bringing a process to a state of statistical control because this is the first step in improving the design of an existing process. Process design improvement involves a reduction in the variation due to chronic problems; and it is difficult, if not impossible, to discover remedies for chronic problems when the process is unstable.

1.7 OFF-LINE QUALITY CONTROL

A product's (or process') performance variation can be reduced by exploiting the nonlinear effects of the product (or process) parameters on the performance characteristics. Because of the importance of product and process designs, quality

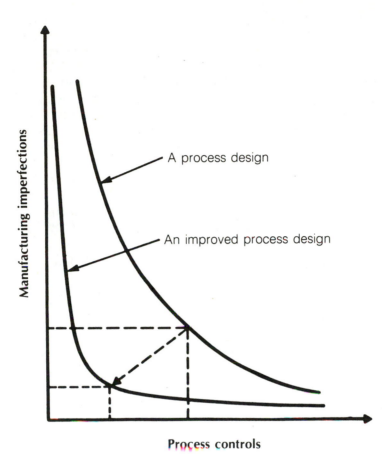

Figure 1.2. Improved Process Design Reduces Both Manufacturing Imperfections and the Need for Process Controls

control must begin with the first step in the product development cycle and it must continue through all subsequent steps. *Off-line quality control methods* are technical aids for quality and cost control in product and the process design. In contrast, *on-line quality control methods* are technical aids for quality and cost control in manufacturing. Off-line quality control methods are used to improve product quality and manufacturability, and to reduce product development, manufacturing, and lifetime costs. The word quality control, as used here, has a very broad meaning. It includes quality planning and quality improvement.

Effective quality control requires off-line quality control methods that focus on quality improvement rather than quality evaluation. Some examples of off-line quality control methods are: sensitivity tests, prototype tests, accelerated life tests, and reliability tests. However, the primary function of these methods is quality evaluation. They are like a doctor's thermometer: a thermometer can indicate that a patient's temperature is too high, but it is not a medication. For example, most reliability tests are primarily concerned with reliability estimation. Reliability estimation is important. For example, good reliability estimates are necessary for quality assurance and for scheduling preventive maintenance. But more often improvement of product reliability is the real issue. In such cases, approximate reliability estimates often suffice and greater emphasis on reliability improvement is needed.

As with performance characteristics, all specifications of product (and process) parameters should be stated in terms of ideal values and tolerances around these ideal values. It is a widespread practice in industry to state these specifications in terms of tolerance intervals only. But this practice can sometimes mislead a manufacturer into producing products whose parameters are barely inside the tolerance intervals. Such products are likely to be of poor quality. Even if all the parameters of a product are within their tolerance intervals, the product may not perform satisfactorily because of the interdependencies of the parameters. For example, if the size of a car door is near its lower tolerance limit and the size of the door frame near its upper tolerance limit, the door may not close properly. A product performs best when all parameters of the product are at their ideal values. Further, the knowledge of ideal values of product (and process) parameters encourages continuous quality improvement.

Taguchi[4] has introduced a three-step approach to assign nominal values and tolerances to product (and process) parameters: system design, parameter design, and tolerance design.

- *System design* is the process of applying scientific and engineering knowledge to produce a basic functional prototype design. The prototype model defines the initial settings of product (or process) parameters. System design requires an understanding of both the customer's needs and the manufacturing environment. A product cannot satisfy the customer's needs unless it is designed to do so. Similarly, designing for manufacturability requires an understanding of the manufacturing environment.

- *Parameter design* is the process of identifying the settings of product (or process) parameters that reduce the sensitivity of engineering designs to the sources of variation. Consider an electrical circuit. Suppose the performance characteristic of interest is the output voltage of the circuit and its target value is y_0. Assume that the output voltage of the circuit is largely determined by the gain of a transistor X in the circuit, and the circuit designer is at liberty to choose the nominal value of this gain. Suppose also that the effect of transistor gain on the output voltage is nonlinear. (See Figure 1.3.) In order to obtain an output voltage of y_0, the circuit designer can select the nominal value of transistor gain to be x_0. If the actual transistor gain deviates from the nominal value x_0, the output voltage will deviate from y_0. The transistor gain can deviate from x_0 because of manufacturing imperfections in the transistor, deterioration during the circuit's life span, and environmental variables. If the distribution of transistor gain is as shown, the output voltage will have a large variation. One way of reducing the output variation is to use an expensive transistor whose gain has a very tight distribution around x_0. Another way of reducing the output variation is to select a different value of transistor gain. For instance, if the nominal transistor gain is x_1, the output voltage will have a much smaller variance. But the mean value y_1 associated with the transistor gain x_1 is far from the target value y_0. Now suppose there is another component in the circuit, such as a resistor, that has a linear effect on the output voltage and the circuit designer is at liberty to choose the nominal value of this component. The circuit designer can then adjust this component to move the mean value from y_1 to y_0. Adjustment of the mean value of a performance characteristic to its target value is usually a much easier engineering problem than the reduction of performance variation. When the circuit is designed in such a way that the nominal gain of transistor X is x_1, an inexpensive transistor having a wide distribution around x_1 can be used. Of course, this change would not necessarily improve the circuit design if it were accompanied by an increase in the variance of another performance characteristic of the circuit.

The utilization of nonlinear effects of product (or process) parameters on the performance characteristics to reduce the sensitivity of engineering designs to the sources of variation is the essence of parameter design. Because parameter design reduces performance variation by reducing the influence of the sources of variation rather than by controlling them, it is a very cost-effective technique for improving engineering designs.

The concept of parameter design is not new to many circuit designers. In the words of Brayton, Hachtel, and Sangiovanni-Vincentelli[5] "various influences, manufacturing, environmental, and aging will cause statistical variations among circuits which are nominally the same. Designers are driven by economical considerations to provide designs which are tolerant to statistical variations."

The essence of parameter design is also familiar to many agronomists. As indicated by Eberhart and Russell,[6] many agronomists routinely conduct experiments to identify plant varieties that can tolerate a wide range of soil, moisture, and weather conditions.

- *Tolerance design* is the process of determining tolerances around the nominal settings identified by parameter design. It is a common practice in industry to assign tolerances by convention rather than scientifically. Tolerances that are too narrow increase the manufacturing cost and tolerances that are too wide increase performance variation. Therefore tolerance design involves a trade off between the customer's loss due to performance variation and the increase in manufacturing cost. Further research is required to better understand Taguchi's approach to tolerance design.

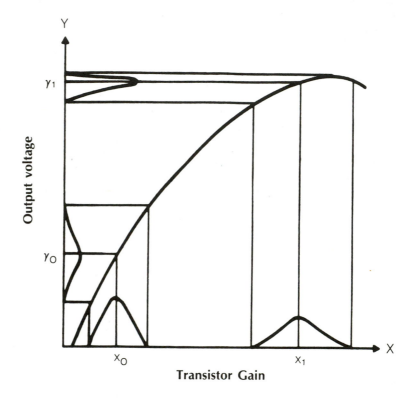

Figure 1.3. Effect of Transistor Gain on the Output Voltage

1.8 PARAMETER DESIGN EXPERIMENTS

Statistically planned experiments can be used to identify the settings of product (and process) parameters that reduce performance variation. Taguchi has proposed a novel approach to the use of statistically planned experiments for parameter design.[4] He classifies the variables that affect the performance characteristics of a product (or process) into two categories: design parameters[1] and sources of noise. The *design parameters* are those product (or process) parameters whose nominal settings can be chosen by the responsible engineer. The nominal settings of design parameters define a product (or process) design specification and vice versa. The *sources of noise* are all those variables that cause the performance characteristics to deviate from their target values. (The sources of noise include the deviations of the actual values of design parameters from their nominal settings.) However, not all sources of noise can be included in a parameter design experiment; limitations are lack of knowledge and physical constraints. The *noise factors*[2] are those sources of noise and their surrogates that can be systematically varied in a parameter design experiment. The key noise factors—those that represent the major sources of noise affecting a product's performance in the field and a process' performance in the manufacturing environment—should be identified and included in the experiment.

The object of the experiment is to identify the settings of design parameters at which the effect of noise factors on the performance characteristic is minimum. These settings are predicted by (1) systematically varying the settings of design parameters in the experiment and (2) comparing the effect of noise factors for each test run.

Taguchi-type parameter design experiments consist of two parts: a design parameter matrix[3] and a noise factor matrix.[4] The *design parameter matrix* specifies the test settings of design parameters. Its columns represent the design parameters and its rows represent different combinations of test settings. The *noise factor matrix* specifies the test levels of noise factors. Its columns represent the noise factors and its rows represent different combinations of noise levels. The complete experiment consists of a combination of the design parameter and the noise factor matrices. (See Figure 1.4.) Each test run of the design parameter matrix is crossed with all rows of the noise factor matrix, so that in the example in Figure 1.4, there are four trials in each test run—one for each combination of noise levels in the noise factor matrix. The performance characteristic is evaluated for each of the four trials in each of the nine test runs. Thus the variation in multiple values of the performance characteristic mimics the product (or process) performance variation at the given design parameter settings.

In the case of continuous performance characteristics (as shown in Figure 1.4), multiple observations from each test run of the design parameter matrix are used

---

1. Glossary equivalent is *control parameter*.
2. Glossary equivalent is *noise parameter*.
3. Glossary equivalent is *control array*.
4. Glossary equivalent is *noise array*.

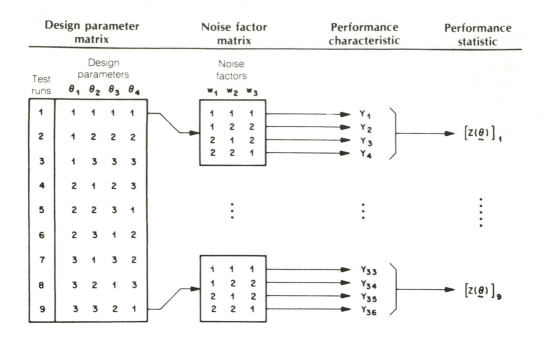

Figure 1.4. An Example of a Taguchi-Type Parameter Design Experiment

to compute a criterion called *performance statistic*. A performance statistic estimates the effect of noise factors. (Examples of performance statistics are provided later in this section.) The computed values of a performance statistic are used to predict better settings of the design parameters. The prediction is subsequently verified by a confirmation experiment. The initial design parameter settings are not changed unless the veracity of the prediction has been verified. Several iterations of such parameter design experiments may be required to identify the design parameter setting at which the effect of noise factors is sufficiently small.

Parameter design experiments can be done in one of two ways: through physical experiments or through computer simulation trials. These experiments can be done with a computer when the function $Y = f(\Theta, \omega)$—relating performance characteristic Y to design parameters Θ, and noise factors ω—can be numerically evaluated.

Taguchi recommends the use of "orthogonal arrays" for constructing the design parameter and the noise factor matrices. All common factorial and fractional factorial plans of experiments are orthogonal arrays. But not all orthogonal arrays are common fractional factorial plans. Kackar,[7] with discussions and response, and Hunter[8] have discussed the use of orthogonal arrays from the statistical viewpoint.

Taguchi recommends the use of criteria he calls "signal-to-noise (s/n)-ratios" as performance statistics. For (non-negative) continuous performance characteristics with fixed target, he has defined three s/n ratios depending on the following three forms of the loss function: the smaller the better, the larger the better, and a specific target value is the best. Suppose $y_1, y_2, ..., y_n$ represent multiple values of the performance characteristic Y. The Taguchi s/n-ratios, denoted here by $Z(\Theta)$, can then be written as follows:

The smaller the better:

$$Z(\Theta) = -10 \, log(\frac{1}{n} \sum y_i^2).$$

The larger the better:

$$Z(\Theta) = -10 \, log(\frac{1}{n} \sum \frac{1}{y_i^2}).$$

A specific target value is the best:

$$Z(\Theta) = 10 \, log(\frac{\bar{y}^2}{s^2}),$$

where

$$\bar{y} = \frac{1}{n} \sum y_i$$

and

$$s^2 = \frac{1}{n-1} \sum (y_i - \bar{y})^2.$$

Kackar[7] has explained these s/n-ratios in more detail. Kackar[7] and Leon, Shoemaker and Kackar[9] have shown how these and other s/n-ratios are related to the quadratic loss function. According to Box,[10] it is better to study the mean $\bar{y}$ and the variance s^2 separately rather than combine them into a single s/n-ratio as Taguchi suggests. Pignatiello and Ramberg[11] have also arrived at the same conclusion after thoroughly investigating a case study.

Taguchi recommends "accumulation analysis" for the data analysis of ordered categorical measurements. Nair[12] and Hamada and Wu[13] have criticized the accumulation analysis method, and they have discussed alternate data analysis techniques.

When the performance characteristic is measured on a binary scale, such as good or bad, Taguchi recommends the following performance statistic:

$$Z(\Theta) = 10 \, log\,(p/(1-p)),$$

where p is proportion good. Leon, Shoemaker, and Kackar[9] have described a reasoning that leads to this performance statistic.

Taguchi has stimulated a great interest in the application of statistically planned experiments to industrial product and process designs. Although the use of statistically planned experiments to improve industrial products and processes is not new (in fact, Tippett[14] used statistically planned experiments in textile and wool industries more than 50 years ago), Taguchi has acquainted us with the breadth of the scope of statistically planned experiments for off-line quality control. From his writings it is clear that he uses statistically planned industrial experiments for at least four different purposes:

1. To identify the settings of design parameters at which the effect of the sources of noise on the performance characteristic is minimum.

2. To identify the settings of design parameters that reduce cost without hurting quality.

3. To identify those design parameters that have a large influence on the mean value of the performance characteristic but have no effect on its variation. Such parameters can be used to adjust the mean value.

4. To identify those design parameters that have no detectable influence on the performance characteristics. The tolerances on such parameters can be relaxed.

AT&T Bell Laboratories experience with Taguchi's ideas are described in Kackar and Shoemaker;[15] Katz and Phadke;[16] Lin and Kackar;[17] Pao, Phadke, and Sherrerd;[18] Phadke;[19] Phadke and others.[20]

Taguchi's applications demonstrate that he has great insight into industrial quality and cost control problems and the efficacy of statistically planned experiments. This insight helps explain Taguchi's impact on the total quality control programs of many Japanese companies. In my opinion, other reasons for the attention he is receiving are as follows:

- He has identified very important quality and productivity problems in product and process designs.

- He has proposed an integrated strategy (involving quality engineering ideas and statistical methods) for attacking these problems.

- He and his associates have successfully taught this strategy to thousands of engineers from many different companies.

- Many Japanese companies have experienced success with Taguchi's strategy for off-line quality control.

- Taguchi's strategy is not used in isolation. It is an integral part of the whole total quality control program. Since many checks and balances are built in a TQC program, the lack of sophistication in some of the statistical methods proposed by Taguchi seems to have no adverse consequences.

It is clear from the October 1985 issue of the *Journal of Quality Technology* and other recent publications that some of Taguchi's ideas are controversial. Most of these controversies are about the sophistication of the statistical methods proposed by Taguchi. These controversies seem to be of secondary importance. We should learn the wide implications of Taguchi's ideas and combine them with the best engineering and statistical techniques we know. Taguchi's quality philosophy is like a soup: it has a few uncooked vegetable pieces, but it is wholesome nonetheless.

ACKNOWLEDGMENTS

The following individuals provided many useful suggestions on the first draft of the paper: Bob Easterling, Gerry Hahn, Ramon Leon, P. B. Narayan, Peter Nelson, Anne Shoemaker, and V. Visvanathan.

REFERENCES

1. Garvin, David A. 1984. "What Does 'Product Quality' Really Mean?" *Sloan Management Review* (Fall):25-43.

2. Taguchi, Genichi, and Yu-In Wu. 1980. *Introduction to Off-line Quality Control Systems*. Nagoya, Japan: Central Japan Quality Control Association. Available from American Supplier Institute, Dearborn, Michigan.

3. Shewhart, Walter A. 1931. *Economic Control of Quality of Manufactured Product.* New York: D. Van Nostrand Company. 1981. Reprint. American Society for Quality Control.

4. Taguchi, Genichi. 1976. *Experimental Designs*. 3d ed. Vols. 1 and 2. Tokyo: Maruzen Publishing Company. (Japanese)

5. Brayton, Robert K., Gary D. Hachtel, and Alberto L. Sangiovanni-Vincentelli. 1981. "A Survey of Optimization Techniques for Integrated-Circuit Design." *Proceedings of the IEEE* **69**:1334-1362.

6. Eberhard, S. A., and W. A. Russell. 1966. "Stability Parameters for Comparing Varieties." *Crop Science* 6 (Jan.-Feb.):36-40.

7. Kackar, Raghu N. 1985. "Off-line Quality Control, Parameter Design, and the Taguchi Method." *Journal of Quality Technology* 17 (Oct. 1985):176-209.
(Quality Control, Robust Design, and the Taguchi Method; Article 4)

8. Hunter, J. Stuart. 1985. "Statistical Design Applied to Product Design." *Journal of Quality Technology* (Oct.):210-221.

9. Leon, Ramon V., Anne C. Shoemaker, and Raghu N. Kackar. 1987. "Performance Measures Independent of Adjustment: An Explanation and Extension of Taguchi's Signal-to-Noise Ratios." *Technometrics* vol. 29, no. 3 (Aug.):253-265.
(Quality Control, Robust Design, and the Taguchi Method; Article 12)

10. Box, George E. P. 1986. *Studies in Quality Improvement: Signal-to-Noise Ratio, Performance Criteria, and Statistical Analysis Part I and II.* (Part II written with Jose Ramirez.) Technical Report. Madison, Wis.: University of Wisconsin, Center for Quality and Productivity Improvement.

11. Pignatiello, Joseph J., Jr., and John S. Ramberg. 1985. "Discussion of Off-line Quality Control, Parameter Design and the Taguchi Method." *Journal of Quality Technology* (Oct.):198-206.

12. Nair, Vijayan N. 1986. "Testing in Industrial Experiments with Ordered Categorical Data." *Technometrics* vol. 28, no. 4 (Nov.):283-291.
(Quality Control, Robust Design, and the Taguchi Method; Article 11)

13. Hamada, Michael, and C. F. Jeff Wu. 1986. *A Critical Look at Accumulation Analysis and Related Methods.* Technical Report. Madison, Wis.: University of Wisconsin, Department of Statistics.

14. Tippett, L. H. C. 1934. "Applications of Statistical Methods to the Control of Quality in Industrial Production." *Journal of the Manchester Statistical Society.*

15. Kackar, Raghu N., and Anne C. Shoemaker. 1986. "Robust Design: A Cost-effective Method for Improving Manufacturing Processes." *AT&T Technical Journal* vol. 65, no. 2 (Mar.-Apr.):39-50.
(Quality Control, Robust Design, and the Taguchi Method; Article 8)

16. Katz, Lewis E., and Madhav S. Phadke. 1985. "Macro-quality with Micro-money." *AT&T Bell Laboratories Record* (Nov.):22-28.
(Quality Control, Robust Design, and the Taguchi Method; Article 2)

17. Lin, Kon M., and Raghu N. Kackar. 1985. "Optimizing the Wave Soldering Process." *Electronic Packaging and Production* (Feb.):108-115.

(Quality Control, Robust Design, and the Taguchi Method; Article 7)

18. Pao, T. W., Madhav S. Phadke, and Chris S. Sherrerd. 1985. "Computer Response Time Optimization Using Orthogonal Array Experiments." *Proceedings of IEEE International Conference on Communications.* IEEE Communications Society (June):890-895.

19. Phadke, Madhav S. 1986. "Design Optimization Case Studies." *AT&T Technical Journal* vol. 65, no. 2 (Mar.-Apr.):51-68.

(Quality Control, Robust Design, and the Taguchi Method; Article 10)

20. Phadke, Madhav S., Raghu N. Kackar, Donald V. Speeney, and Mike J. Grieco. 1983. "Off-line Quality Control in Integrated Circuit Fabrication Using Experimental Design." *The Bell System Technical Journal* vol. 62, no. 5 (May-June):1273-1309.

(Quality Control, Robust Design, and the Taguchi Method; Article 6)

2

MACRO-QUALITY WITH MICRO-MONEY

Lewis E. Katz and
Madhav S. Phadke

2.1 INTRODUCTION

Small though it is, a microchip requires a large number of specifications to govern its design, manufacturing, and operation.

The manufacturing process is especially critical, since it's caught between a rock and a hard place. On the one hand, the design specs set up difficult-to-maintain line widths and device features. On the other hand, the operating specs require a high level of reliability.

A conventional approach to microchip fabrication specifies ultra-precise machines that maintain astonishingly tight tolerances. But narrow tolerances are always expensive, whether they involve materials or machinery.

In microelectronics production, the company must install equipment capable, for example, of precisely controlling the temperatures and the voltages used in processing the microchip wafers. Otherwise, the plant will have to throw away a lot of unacceptable product in order to deliver a given number of microchips.

However, if the manufacturing process can be designed to broaden the tolerance range—while still producing the same high-quality product—everybody benefits. The rewards are less waste, faster production, lower costs, and a better competitive position in the marketplace.

It's one thing to say that the process should be adjusted to ease the handicaps of narrow tolerances in each manufacturing step. But it's something else again to actually accomplish it and still keep the quality-level needed in microelectronics devices.

Manufacturing process engineers normally have to approach the production of a new chip cautiously. There is a lot of synergy or interrelationship between each of the many process steps. Test after test after test must be performed with computer simulation—as well as through hardware experimentation—by the device engineers, to determine the parameters for each process step (i.e., dose, times and temperatures of processing, etc.).

Optimizing each process step has been traditionally accomplished by the experience and judgment of the process engineers. Now, however, a new mathematical procedure, known as the "robust design method," eliminates a great deal of the guesswork and tedious test procedures needed in the past.

Robust design requires close cooperation between the quality assurance people and the manufacturing process engineers. Proof that the approach works is evident in the latest major project that used robust design to speed up the evolution of the manufacturing process.

The project: AT&T's megabit memory chip, now entering the manufacturing phase before any other company's megabit chip.

There are two stages in implementing the robust design. First, the device must be made as insensitive as possible to fluctuations in manufacturing processes. Second, the right tolerance level must be found within the economics of the product.

Suppose it costs a hundred dollars to reduce the process tolerance by 50 percent, and it leads to more than a hundred dollars worth of quality improvement (as measured by fewer defective products and closer-to-target performance). Then the narrower tolerance should be specified. But if the resulting quality improvement is worth *less* than a hundred dollars, we should *not* specify the narrower tolerance. This is the common-sense principle behind the second stage of robust design—and it can often lead to much larger benefits compared to the investment.

Meanwhile, in the first stage of robust design we find new center values for process parameters that obtain hundreds of dollars worth of quality improvement—without narrowing the tolerances. This is the benefit of the new emphasis brought by robust design: make the manufacturing process insensitive to fluctuations.

Because the marketplace is highly competitive in microelectronics, this work in cost-effectiveness can mean the difference between profit and loss. And because robust design often means that a new product—such as the megabit chip—can reach the marketplace before the competition does, this concept can literally mean the difference between survival and demise.

2.2 ROBUST DESIGN VS. AUDITING

In the past, the quality assurance approach mainly focused on auditing the quality of the product after it was manufactured—that is, between the factory shipping point and when it actually reached the customer.

But today quality assurance is taking on a more active function: how to do it right the first time. This means that the audit after fabrication becomes less important, since the quality is far more consistent. (See *New initiatives in quality assurance*, AT&T Bell Laboratories RECORD, December 1983.)

The statistical control approach of Walter Shewhart 60 years ago used a principle aimed at keeping the manufacturing process in control. That is, an engineer states that "this process bath temperature should be 100 degrees F." The Shewhart process control charts set up diagnoses to report any deviation from 100 degrees, flagging the deviation so the bath can be brought back to 100 degrees.

Robust design works differently. In this concept, the engineers search for the best design for the process so that the bath will not deviate far from 100 degrees, even when something goes wrong. In effect, robust design starts one step before the Shewhart control charts.

The process engineer must consider six or eight variables—or even more—that can affect the process step. The inter-relationships between these variables are not necessarily linear.

Assume that a given two-step process has four variables in the first step and five variables in the second step. In order to study all possible combinations of the settings of these variables, many thousands of tests would have to be conducted in order to determine which combination is the best for the process. See Table 2.1 for an example.

TABLE 2.1. VARIABLES—MIXED HALOGEN PROCESS*

Step 1	Level 1**	Level 2	Level 3
CHAMBER PRESSURE (μm)	1	1.8	
FLOW RATE (TOTAL) (SCCM)	1	1.5	
ADDITIVE (% TOTAL FLOW)	1	2	3
DC BIAS (VOLTS)	1	1.1	1.2
Step 2			
CHAMBER PRESSURE (μm)	1	2	3
FLOW RATE (TOTAL (SCCM)	1	1.75	2.5
ADDITIVE (%)	1	15	30
DC BIAS (VOLTS)	1	1.25	
OVERETCH (A)	1	1.5	2.0

*   This chart shows four variables selected as influencing the first step, and five variables for the second step, of a mixed halogen process used in microelectronics manufacturing. Each of these variables has two or three possible settings called "levels." These levels have been normalized so that the first level is numerically equal to one.

**   All values normalized to 1.

Of course, this assumes that the process engineer has identified all the right variables. But since the microelectronics field is evolving at a forest-fire rate, the engineer often is working in unexplored territory. This could result in a lot of wasted effort, if many tests are conducted with the wrong variables.

This is where the robust design concept pays off. Instead of laboriously running hundreds of physical tests—often sandwiching them between production runs that share the line—the process engineer supplies the pertinent variables to the quality assurance center, where a small number of judiciously chosen process conditions are organized into a chart-like matrix using a technique called the orthogonal array experimentation method.

Experimental data from this small number of process conditions—together with a special data analysis method—can give much information on eight or even more process variables. In most cases, this limited quantity of data is sufficient to allow the pinpointing of the best operating levels for these process variables.

If the tests don't produce results as good as expected, the process engineer must look for more control variables.[1] These might be humidity, or electrode voltage, or gas pressure, or some other aspect. This is where judgment comes into play, so the process deviations can be tamed by identifying the right control variable.

2.3 THE ORTHOGONAL ARRAY METHOD

The traditional method of studying a large number of variables has been to review them one at a time. This is obviously an inefficient and costly approach.

Grouping the variables into columns that form a matrix called an "orthogonal array," the quality assurance engineer can extract much more precise information than if the experiments were conducted with a single-factor by single-factor approach. In every pair of orthogonal columns (see Figure 2.1), all combinations of different variable levels occur, and they occur the same number of times.

In the orthogonal array method, many factors are studied simultaneously. In successive tests, the engineers change the values of many variables. A suitable orthogonal array is used to determine how these values should be changed, so that the effect of each variable can be separated from the others at the end of the study.

Once the robust design programs have determined the ideal levels of the variables, the process engineers can run a series of confirmation experiments. The robust design concept does not eliminate experiments—it simply makes the selection of the process variable settings and the production tolerances much more efficient, saving time and costs.

---

1. Glossary equivalent is *control parameter*.

The L_{18} Orthogonal Array

Experiment Number	Column Number & Factor							
	1 A	2 BD	3 C	4 E	5 F	6 G	7 H	8 I
1	1	1	1	1	1	1	1	1
2	1	1	2	2	2	2	2	2
3	1	1	3	3	3	3	3	3
4	1	2	1	1	2	2	3	3
5	1	2	2	2	3	3	1	1
6	1	2	3	3	1	1	2	2
7	1	3	1	2	1	3	2	3
8	1	3	2	3	2	1	3	1
9	1	3	3	1	3	2	1	2
10	2	1	1	3	3	2	2	1
11	2	1	2	1	1	3	3	2
12	2	1	3	2	2	1	1	3
13	2	2	1	2	3	1	3	2
14	2	2	2	3	1	2	1	3
15	2	2	3	1	2	3	2	1
16	2	3	1	3	2	3	1	2
17	2	3	2	1	3	1	2	3
18	2	3	3	2	1	2	3	1

Figure 2.1. The Orthogonal Array. This orthogonal array is a matrix of 18 experiments in eight variables. Each row specifies the level of every variable to be used in that experiment. The columns of the arrays are pairwise orthogonal; i.e., in every pair of columns, all combinations of variable levels occur, and they occur an equal number of times.

2.4 A QUANTUM JUMP

Robust design was first implemented at AT&T Bell Laboratories in 1980. Some portions of the concept existed prior to that time—but the present form is a quantum jump over what was used before.

Originally known as off-line quality control, the method evolved in Japan after World War II. It was developed by Professor Genichi Taguchi, former director of the Japanese Academy of Quality and a recipient of the Deming award. (He is now a consultant to AT&T and other companies.) The AT&T work in this field has been built on his findings, but has since gone further.

Off-line quality control is used routinely by many leading Japanese industries to produce high-quality products at low cost. Professor Taguchi's work was stimulated by an American, W. Edwards Deming, who alerted Japanese management to the benefits of applying quality control on the production line rather than on finished goods.

In today's plants, fabrication of integrated circuits is a complex and lengthy process. The first AT&T application of what is now called robust design was to reduce the variance of "contact window" size—tiny holes with diameters of about 3.5 micrometers that are etched through oxide layers about two micrometers thick. An AT&T WE® 32100 microprocessor chip, for example, has about 250,000 windows in an area about 1.5 by 1.5 centimeters.

Only 18 different process conditions were studied to determine the optimum variable levels of nine control variables from the possible 6000 combinations. These levels were subsequently used in fabricating both the WE 32100 microprocessor and the WE 4000 microcomputer, as well as certain other microchips under development at the time.

The next major application of robust design was in the development of fabrication processes for the AT&T WE 256K memory chip. The development process for the 256K DRAMs was sequential—that is, the design of the chip was followed by the design of the manufacturing process.

The megabit memory chip, however, was market-driven from the start. Development of the fabrication process was parallel with development of the chip, so the plant was ready to "pounce" when the design was completed.

Just as the robust design concept of today is a quantum jump from the earlier methods of quality assurance, the demands made on the process engineers are quantum jumps from previous technology.

A decade ago, the pattern lines for very-large-scale-integration (VLSI) circuits were so wide the wafer could be immersed in a wet bath for etching. But wet-etching acts both horizontally and vertically—isotropic etching—so there was some undercutting unless the mask size was adjusted to compensate for it.

Now the pattern lines are so narrow that adjustment of the mask would completely eliminate them. So the new fabrication processes use dry etching, working with a plasma atmosphere. This involves a reduced pressure in a housing, with a certain chemical reaction.

When the product designers come to the process engineers and ask if they can shrink certain sizes, or do something else to a design, the engineer now knows so much more about the process that an instant answer is possible—"Of course we can!" or "Forget it!"

Some major steps in the fabrication of VLSI circuits include source-and-drain formation, oxide growth, patterning of polysilicon and tantalum silicide, deposition of baron-phosphorus-silicon glass, window photolithography, patterning of aluminum, protective coating, and packaging. Each of these steps has a variety of sub-stages, for which there are numerous parameters with several variation levels.

Tantalum-silicide etching is only one of those major steps, but it's considered to be the most critical. The etching machine employed on the production line recently was yielding only two-thirds of its rated capacity because of non-uniform etching. The problem posed by this situation offered three choices: replace the machine with another make, add more machines of the same make, or fix the problem with improved or new chemistry.

Conventional analysis methods required a prohibitively large number of experiments to investigate all the parameters. But robust design analysis used matrix experiments to achieve uniform etching within the deadline period (20 days). A modified process was introduced into the existing machines, increasing their output by 50 percent.

The robust design approach saved the plant $1.2 million in equipment replacement costs plus the expense of disruption on the factory floor. And it optimized the tantalum-silicide etching process.

Other successful applications of robust design programs within AT&T include a reduction of response time for the UNIX^{TM} operating system by a factor of three, a four-time increase in life for router bits, a 65 percent improvement in eliminating visual defects in aluminum etching of microchips, and several implementations within computer aided circuit design systems.

2.5 TESTING—ONE, TWO...

Physical experiments are still needed in setting up a fabrication process for a new microchip. Robust design helps to minimize the number of experiments.

Suppose a given etching step has nine variables, and each variable offers two or three choices. Some 6000 tests would be required to prove out all these combinations—but nobody can afford that many tests. So the robust design team chooses 18 of those 6000 tests, and comes up very close to the best solution. Using the data from all 18 tests, the team can select the optimum combination from the total number of possible combinations.

These tests proceed in quantum steps: the forms are organized as an eight-experiment, 18-experiment, 27-experiment, or 36-experiment form. The selected tests are chosen judiciously, but there is no way to know whether the one that is isolated by the process is the absolutely best process.

What IS known is that the results reported by the robust design program are significantly better (or not) than what was happening before. It's not necessary to be THE best—it's more important to be NEAR the best.

This approach is so effective that even prototype production can be more precise than is usually expected. For example, a design team recently visited the Allentown plant. They wanted megabit chips with line widths that deviated by a certain amount from the specified width, so they could run experiments on how those variations in line width affected the speed of the device.

But the production equipment persisted in turning out chips that were so consistent that the variations were not even a third of what the designers were looking for. Process engineers eventually had to make three separate masks so they would have chips with three separate line widths.

Robust design provides this precision in a cost-effective manner. In fact, cost-effectiveness is a primary achievement of robust design. Without constant efforts to reduce fabrication costs, AT&T products would not be in strong competitive positions.

In this respect, robust design is an important contributor to keeping AT&T robust.

3

QUALITY ENGINEERING USING DESIGN OF EXPERIMENTS

M. S. Phadke

3.1 INTRODUCTION

The Webster's dictionary defines quality control as "an aggregate of activities (as design analysis and statistical sampling with inspection for defects) designed to ensure adequate quality in manufactured products." This is a passive definition in that it concerns only with 'adequate quality' and makes no mention of an effort to constantly improve the quality. In this paper by quality control we also mean an active effort to improve the quality.

Problems that can arise from an exclusive use of the passive definition of quality control are illustrated by the following example taken from April 17, 1979 issue of a Japanese newspaper "The Asahi."[1] Figure 3.1 shows the distribution of a quality characteristic, called the color density, of television sets made by Sony-U.S.A. and Sony-Japan. The desired level is m, while m ± 5 define the tolerance limits. In Sony-U.S.A where attention was focused on the 'adequate quality' defined by the tolerance limits, we see a more or less uniform distribution of the color density with all television sets complying with the specified limits. The uniform distribution is a direct result of the quality control achieved by screening. Televisions made by Sony-Japan, however, have a bell-shaped distribution with a small fraction of televisions lying outside the tolerance limits. This is due to a conscious effort to come as close to m as possible. Televisions which have color density near m are of best quality, i.e., grade A. As one goes away from m in either direction, the grade drops. Sony-Japan is seen to make many more sets of grade A and many fewer sets of grade C when compared to Sony-U.S.A. Suppose the customer's repair cost is proportional to the squared deviation of the color density from the target mean value. The ratio of the expected customer's cost for Sony-U.S.A sets to that for Sony-Japan sets is equal to the ratios of their respective second moments. In this case the cost of Sony-U.S.A. sets is about three times higher than the cost of Sony-Japan sets!

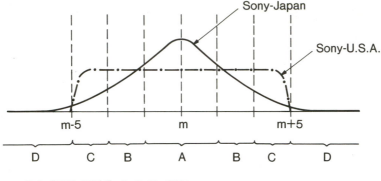

Ref: "THE ASAHI", April 17, 1979

Figure 3.1. Distribution of Color Density in TV Sets

Minimizing the customer's cost of quality is the right approach to quality control. It is needed for a producer to succeed in today's competitive environment. This cannot be accomplished by simply satisfying the tolerance limits. Instead, for every quality characteristic we need to achieve a distribution which is centered at the desired mean value with the smallest economically justifiable variance. Many statistical methods are needed to accomplish good quality control as discussed in standard texts like Duncan.[2] In this paper we will concentrate on the role of design of experiments and analysis of variance.

Professor Genichi Taguchi of Japan, a Deming Award winner and a former Director of the Japanese Academy of Quality visited Bell Laboratories in August through October of 1980. During his visit he gave several lectures on the Japanese methods of applying design of experiments and analysis of variance (ANOVA) in quality control. Our experience with his methods over the last two years indicates that his methods have a high potential in improving quality and productivity.

This paper is divided into six sections. In Section 3.2 we will give a brief perspective of Taguchi's view of quality engineering (Taguchi,[3] and Kackar and Phadke[4]). In Section 3.3 we will describe our experience in improving the window cutting process of integrated circuit fabrication (see Phadke and others[5]). An application in circuit design will be described in Section 3.4 (Phadke and Blaine[6]). Discussion on the design of complex experiments is given in Section 3.5 and concluding remarks are presented in Section 3.6.

3.2 QUALITY ENGINEERING

Professor Genichi Taguchi views the quality of a product from the point of view of its impact on the whole society (see Taguchi[3] and Kackar and Phadke[4]). As a quantitative measure he defines the quality of a product to be the *loss imparted by the product to the society* from the time the product is shipped to the customer. A better quality product is one which imparts smaller loss to the society. There are many ways in which a product can cause losses to the society. Here, we concentrate on losses due to deviation of the product's *functional characteristic* from its desired *target value*. Such losses are called losses due to functional variation. For example, the output voltage of a power supply circuit may be the functional characteristic of interest. The desired value of the output voltage is then, the target value. Any losses due to deviation of the output voltage from its desired value are losses due to functional variation.

The uncontrollable factors which cause the functional characteristic of a product to deviate from its target value are called noise factors.[1] The noise factors can be classified into three categories: 1) external factors, 2) manufacturing imperfections and 3) product deterioration. The variations in the operating environment like temperature, humidity, vibrations due to nearby machinery, supply voltage etc., and human errors in operating the product are the external noise factors. By manufacturing imperfection we mean the variation in the product parameters from unit to unit which is inevitable in a manufacturing process. An example of the manufacturing imperfection is that the resistance of a particular resistor in an amplifier may be specified to be 100 kilo-ohms but in a particular unit it turns out to be 101.5 kilo-ohms. These manufacturing variations are a measure of manufacturing capability. Increase in the resistance of a resistor with age, loss of resilience of springs or wearing out of the parts of a motor due to friction are examples of noise sources due to product deterioration.

The overall aim of quality engineering is to produce products which are *robust* with respect to all noise factors. Robustness implies that the product's functional characteristic is not sensitive to variations in the noise factors. To achieve robustness quality control efforts have to begin during the product design and continue on through production process design and actual manufacturing. During product design one can take care of all three categories of noise factors while during production process design and actual manufacturing one can only take care of manufacturing imperfections. The quality control efforts during product design and production process design are called *off-line quality control* while the efforts during the actual production are called *on-line quality control*.

---

1. Glossary equivalents are *noise* or *noise parameter*.

In off-line quality control, Professor Taguchi's overall approach is to follow a two-step procedure in preference to approaches which involve finding and controlling noise factors.

1. Determine levels of manipulatable factors at which the effect of noise factors on the functional characteristic is minimum. This is called *parameter design*.

2. If the reduction in the variation of the functional characteristic achieved by parameter design is not enough then as a last alternative, narrower tolerances are specified for the noise factors. This is called *tolerance design*.

These procedures have been extensively used by many Japanese industries for improving the quality of their products while simultaneously controlling the cost.

The main statistical techniques used in parameter design are design of experiments and analysis of variance. In the following sections we will illustrate the use of these methods.

As an on-line technique for improving quality and yield, Professor G. E. P. Box[7] has suggested the use of experimental design through "evolutionary operations," commonly known as "EVOP." While permitting us to manufacture products under current capability, EVOP helps gather useful information for making improvements in the process.

3.3 OFF-LINE QUALITY CONTROL FOR THE INTEGRATED CIRCUIT FABRICATION PROCESS

The application described in this section is based on the work done by Phadke, Kackar, Speeney and Grieco[5] in collaboration with G. Taguchi. A typical very large scale integrated circuit (IC) chip has thousands of contact windows, most of which are not redundant. It is critically important to produce windows of size very near the target dimension. Photolithography is used to form or cut these windows, which are about 3 μm in size, in an oxide layer of about 2 μm thickness. There are nine factors associated with the process: A) Mask dimension, B) Viscosity of the photoresist, C) Spin speed, D) Bake temperature, E) Bake time, F) Aperture, G) Exposure time, H) Development time, and I) Etch time.

In August 1980 when we started the study, the window size had a large variance and in many cases the windows were not even open. Our aim in this study was to find the best levels for each of the nine factors so that window size would have the smallest variance, while keeping the mean at the target value of 3 μm. Note that in addition to the mean we are also interested in the variance.

The potential levels for each of the nine factors are listed in Table 3.1. The standard levels of these factors prior to August 1980 are also shown in the table.

Factors A, B and D have two levels each while the remaining six factors have three levels each. A complete factorial design in this case is very large, $3^6 \times 2^3 = 5832$ experiments.

TABLE 3.1. TEST LEVELS

Label	Factors Name		Levels	
A	Mask Dimension (μm)		2	2.5
B	Viscosity		204	206
C	Spin Speed	Low	Normal	High
D	Bake Temperature	90°C	105°C	
E	Bake Time (Min)	20	30	40
F	Aperture	1	2	3
G	Exposure Time	+20%	Normal	-20%
H	Develop. Time (Sec)	30	45	60
I	Plasma Etch Time (Min)	14.5	13.2	15.8

Standard
Levels

Dependence of Spin Speed on Viscosity

		Spin Speed (rpm)		
		Low	Normal	High
Viscosity	204	2000	3000	4000
	206	3000	4000	5000

Dependence of Exposure on Aperture

		Exposure (PEP-Setting)		
		20% Over	Normal	20% Under
Aperture	1	96	120	144
	2	72	90	108
	3	40	50	60

3.3.1 ORTHOGONAL ARRAY DESIGN

The fractional factorial design used for this study is given in Table 3.2. It is the L_{18} orthogonal array design consisting of 18 experiments taken from Taguchi and Wu.[8] The rows of the array represent runs while the columns represent the factors. Here we treat BD as a joint factor with the levels 1, 2 and 3 representing the combinations B_1D_1, B_2D_1, and B_1D_2, respectively.

TABLE 3.2. L_{18} ORTHOGONAL ARRAY

Experiment Number	Factor							
	A	BD	C	F	F	G	H	I
1	1	1	1	1	1	1	1	1
2	1	1	2	2	2	2	2	2
3	1	1	3	3	3	3	3	3
4	1	2	1	1	2	2	3	3
5	1	2	2	2	3	3	1	1
6	1	2	3	3	1	1	2	2
7	1	3	1	2	1	3	2	3
8	1	3	2	3	2	1	3	1
9	1	3	3	1	3	2	1	2
10	2	1	1	3	3	2	2	1
11	2	1	2	1	1	3	3	2
12	2	1	3	2	2	1	1	3
13	2	2	1	2	3	1	3	2
14	2	2	2	3	1	2	1	3
15	2	2	3	1	2	3	2	1
16	2	3	1	3	2	3	1	2
17	2	3	2	1	3	1	2	3
18	2	3	3	2	1	2	3	1

Here are some of the properties and considerations of this design:

(i)   This is a main effects only design, i.e., the response is approximated by a separable function.

(ii)   The columns of the array are pairwise orthogonal. That is, in every pair of columns, all combinations of levels occur and they occur equal number of times.

(iii)   Consequently, the estimates of the main effects of all factors and their associated sum of squares are independent under the assumption of normality and equality of observation variance. So the significance tests for these factors are independent.

(iv)   The estimates of the main effects can be used to predict the response for any combination of the parameter levels. A desirable feature of this design is that the variance of the prediction error is the same for all such parameter level combinations covered by the full factorial design.

(v)   Some of the factors (B, D, F and G) are discrete while the others are continuous. For the continuous factors, the main effects of the three level factors can be broken down into linear and quadratic terms using the Chebychev orthogonal polynomials. For the two level factors we can only estimate the linear effects. So this design can be used to also estimate the response surface with respect to the continuous factors.

(vi)   It is known that the main effect only models are liable to give misleading conclusions in the presence of interactions. However, in the beginning stages of a study the interactions are assumed to play a smaller role.

An attempt to study all two factor interactions in the beginning stages of a study can be compared with a geologist who drills 100 holes in a square mile area in search of oil rather than drilling 100 holes distributed over a 100 square mile area. If we wished to study all two factor interactions, with no more than 18 experiments we would have enough degrees of freedom for studying only two three-level factors, or five two-level factors! That would mean in the present study we would have to eliminate half of the process factors without any experimental evidence.

Orthogonal array designs can, of course, be used to study interactions. We will consider an example of that in Section 3.5.

In conducting experiments of this kind, it is common for some wafers to get damaged or break. Also, the wafer-to-wafer variability, which relates to the raw material variability, of window sizes is typically large. So we decided to run each experiment with two wafers. Three quality measures were used: pre-etch line width, post-etch line width and post-etch window size. Five chips were selected from each wafer for making these measurements. These chips correspond to specific locations on a wafer—top, bottom, left, right and center. Thus for every experiment we have 10 readings for each of the three quality characteristics.

3.3.2 DATA ANALYSIS

Both the pre-etch and the post-etch line widths are continuous variables. For each of these variables the statistics of interest are the mean and the standard deviation. The objective of our data analysis is to determine the factor level combination such that the standard deviation is minimum while keeping the mean on target. We will call this factor level combination as the optimum combination. Professor Taguchi's method for obtaining the optimum combination is given next.

Note that minimizing the standard deviation is tantamount to minimizing the influence of the noise factors. In this experiment, the noise factors are the normal variations in the spin speed, the temperature of the baking oven, the thickness of the photoresist across the wafer, etc. These are the factors which cannot be manipulated by the process engineer.

Let us first consider the case where there is only one response variable. Instead of working with the mean and the standard deviation, we work with the transformed variables—the mean and the signal-to-noise ratio (S/N ratio). In the decibel scale, the S/N ratio is defined as:

$$S/N \text{ ratio} = 20 \log_{10} \left(\frac{\text{Mean}}{\text{Standard Deviation}} \right)$$

$$= -20 \log_{10}(\text{coefficient of variation}).$$

In terms of the transformed variables, the optimization problem is to determine the optimum factor levels such that the S/N ratio is maximum while keeping the mean on target. This problem can be solved in two stages as follows:

1. Determine factors which have a significant effect on the S/N ratio. This is done through the analysis of variance (ANOVA) of the S/N ratios. These factors are called the *control factors*,[2] implying that they control the process variability. For each control factor we choose the level with the highest S/N ratio as the optimum level. Thus the overall S/N ratio is maximized. (Note that we are working with the main effects models. When interactions are present, they have to be appropriately accounted for.)

2. Select a factor which has the smallest effect on the S/N ratio among all factors that have a significant effect on the mean. Such a factor is called a *signal factor*.[3] Ideally, the signal factor should have no effect on the S/N ratio. Choose the levels of the remaining factors (factors which are neither control factors nor signal factors) to be the nominal levels prior to the optimization experiment. Then set the level of the signal factor so that the mean response is on target.

Professor Taguchi suggests that in practice the following two aspects should also be considered in selecting the signal factor: a) preferably, the relationship between the mean response and the levels of the signal factor should be linear and b) it should be convenient to change the signal factor during production. These aspects are important from the on-line quality control considerations. Under Professor Taguchi's on-line quality control system the signal factor is used during manufacturing to compensate for small deviations in the mean response.

Why do we work in terms of the S/N ratio rather than the standard deviation? Frequently, as the mean decreases the standard deviation also decreases and vice versa. In such cases, if we work in terms of the standard deviation, the

---

2. Glossary equivalent is *control parameter*.
3. Glossary equivalent is *adjustment parameter*.

optimization cannot be done in two steps, i.e., we cannot minimize the standard deviation first and then bring the mean on target.

Through many applications, Professor Taguchi has empirically found that the two stage optimization procedure involving the S/N ratios indeed gives the parameter level combination where the standard deviation is minimum while keeping the mean on target. This implies that the engineering systems behave in such a way that the manipulatable production factors can be divided into three categories:

1. control factors, which affect process variability as measured by the S/N ratio.

2. signal factors, which do not influence (or have negligible effect on) the S/N ratio but have a significant effect on the mean, and

3. factors which do not affect the S/N ratio or the process mean.

Our two stage procedure also has an advantage over a procedure that directly minimizes the mean square error from the target mean value. In practice, the target mean value may change during the process development. The advantage of the two stage procedure is that for any target mean value (of course, within reasonable bounds) the new optimum factor level combination is obtained by suitably adjusting the level of only the signal factor. This is so because in step 1 of the algorithm the coefficient of variation is minimized for a broad range of mean target values.

When there are multiple responses, engineering judgement is used to resolve any conflicts that may be present among the response variables.

Sometimes the experimental data are categorical or mixed categorical-continuous which was the case with the post-etch window size data. Such data can be effectively analyzed by the "accumulation analysis method" proposed by Taguchi.[9] For the lack of space we will not discuss the method here.

3.3.3 RESULTS

Using the above data analysis method, factors A, B, C, E, F and H were identified as control factors while G was identified as the signal factor. The data suggested that the mask dimension be changed from 2.0 μm to 2.5 μm, spin speed from 3000 rpm to 4000 rpm and development time from 45 seconds to 60 seconds. The exposure time is to be adjusted to get the correct mean value of the line width and the window size. The experiment indicated that the levels of the other factors were optimum to start with.

A verification experiment conducted along with the implementation of the optimum condition showed the following improvements:

i) The standard deviation of the pre-etch window size, which is a standard measure of quality, reduced from 0.29 μm to 0.14 μm.

ii)   The cases of window not open (i.e., window size equal to zero) reduced from 0.12 windows/chip to 0.04 windows/chip. This amounts to a three fold reduction in the defect density.

iii)   Observing these improvements over several weeks, the process engineers gained a confidence in the stability and robustness of the new process parameters. So they eliminated a number of inprocess checks. As a result the overall time spent by the wafers in window photolithography has been reduced by a factor of two.

Thus, this is an example where substantial gains in quality and productivity were achieved through the use of a designed experiment and ANOVA.

3.4 QUALITY AND COST CONTROL IN PRODUCT DESIGN

The first step in designing a product is system design, i.e., arriving at the basic product configuration. In case of circuit design, it means choosing the appropriate circuit diagram. This is mainly a function of a specialist in that technical field. Next, the levels of the component characteristics and their tolerances have to be chosen. Typically, the wider the tolerance, the lower the component cost. Also, a wider tolerance leads to a wider variation in the functional characteristic of the product. This poses the typical dilemma of having to trade quality and cost. There is a way around this. It is to exploit the nonlinearity of the relationship between the component characteristics and the functional characteristic by choosing the nominal levels of the component characteristics so that even with the wide tolerances on the component characteristics, the functional characteristic of the product will have a small variation. If any further reduction in the variance of the product's functional characteristic is to be achieved it can be done by reducing the tolerances on the component characteristics.

We will see how the design of experiments and ANOVA can be used in product design with the help of the design of a differential operational amplifier which is commonly used in telecommunications. This example is taken from Phadke and Blaine.[6] The circuit has two current sources, five transistors and eight resistors. It is expected to function over a wide temperature range. The balancing property of this circuit puts certain restrictions on the values of the circuit parameters. Due to these restrictions, the engineers can independently specify the values of only five circuit parameters. We call these parameters as "design factors," meaning that they are manipulatable parameters. The manufacturing variations in the resistance, the transistors and the current sources lead to an inbalance in the circuit and cause an offset voltage. There are twenty tolerance parameters in this circuit. Thus, together with the temperature we have twenty-one "noise parameters." The objective of the design is to choose the levels of the design parameters such that the mean square offset voltage under the normal manufacturing tolerances is minimum.

In mathematical terms, our problem is to search the five dimensional space spanned by the design factors for a point which has the smallest mean square offset voltage. Design of experiments is a very effective tool in solving this problem. Table 3.3 gives the levels for each of five design parameters which were selected to span the potential design region. The middle levels of each of the factor represents the starting design. Table 3.4 gives the orthogonal array L_{36} taken from Taguchi[10] The factors A, B, C, D, and E were assigned to columns 12, 13, 14, 15, and 16, respectively. The submatrix of the matrix L_{36} formed by columns 12 through 16 is denoted by { I_{ik} } and is referred to as the *control orthogonal array*.

TABLE 3.3. CONTROL FACTORS

| | | | | Levels | |
Label	Name	Description	1	2	3
A	RFM	FEEDBACK RESISTANCE, MINUS TERMINAL (KΩ)	35.5	71	142
B	RPEM	EMITTER RESISTANCE, PNP MINUS TERMINAL (KΩ)	7.5	15	30
C	RNEM	EMITTER RESISTANCE, NPN MINUS TERMINAL (KΩ)	1.25	2.5	5
D	CPCS	COMPLEMENTARY PAIR CURRENT SOURCE (μA)	10	20	40
E	OCS	OUTPUT CURRENT SOURCE (μA)	10	20	40

PRESENT
LEVEL

For ten of the noise parameters two levels were chosen situated at one standard deviation on either side of the mean. These noise parameters were assigned to columns 1 through 10 of the matrix L_{36}. For the remaining eleven noise parameters three levels were chosen, situated at the mean and at $\sqrt{3}/2$ times standard deviation on either side of the mean. These parameters were assigned to columns 12 through 22 of the matrix L_{36}. The submatrix of L_{36} formed by columns 1 through 10 and 12 through 22 is denoted by { J_{jl} } and is referred to as the *noise orthogonal array*. Note that the control and the noise orthogonal arrays do not have to come from the same orthogonal matrix, in this case the L_{36} matrix. Here we choose the control and noise orthogonal arrays in this way for convenience.

TABLE 3.4. L_{36} ORTHOGONAL ARRAY

No.	1	2	3	4	5	6	7	8	9	10	11	12	13	14	15	16	17	18	19	20	21	22	23
1	1	1	1	1	1	1	1	1	1	1	1	1	1	1	1	1	1	1	1	1	1	1	1
2	1	1	1	1	1	1	1	1	1	1	1	2	2	2	2	2	2	2	2	2	2	2	2
3	1	1	1	1	1	1	1	1	1	1	1	3	3	3	3	3	3	3	3	3	3	3	3
4	1	1	1	1	1	2	2	2	2	2	2	1	1	1	1	2	2	2	2	3	3	3	3
5	1	1	1	1	1	2	2	2	2	2	2	2	2	2	2	3	3	3	3	1	1	1	1
6	1	1	1	1	1	2	2	2	2	2	2	3	3	3	3	1	1	1	1	2	2	2	2
7	1	1	2	2	2	1	1	1	2	2	2	1	1	2	3	1	2	3	3	1	2	2	3
8	1	1	2	2	2	1	1	1	2	2	2	2	2	3	1	2	3	1	1	2	3	3	1
9	1	1	2	2	2	1	1	1	2	2	2	3	3	1	2	3	1	2	2	3	1	1	2
10	1	2	1	2	2	1	2	2	1	1	2	1	1	3	2	1	3	2	3	2	1	3	2
11	1	2	1	2	2	1	2	2	1	1	2	2	2	1	3	2	1	3	1	3	2	1	3
12	1	2	1	2	2	1	2	2	1	1	2	3	3	2	1	3	2	1	2	1	3	2	1
13	1	2	2	1	2	2	1	2	1	2	1	1	2	3	1	3	2	1	3	3	2	1	2
14	1	2	2	1	2	2	1	2	1	2	1	2	3	1	2	1	3	2	1	1	3	2	3
15	1	2	2	1	2	2	1	2	1	2	1	3	1	2	3	2	1	3	2	2	1	3	1
16	1	2	2	2	1	2	2	1	2	1	1	1	2	3	2	1	1	3	2	3	3	2	1
17	1	2	2	2	1	2	2	1	2	1	1	2	3	1	3	2	2	1	3	1	1	3	2
18	1	2	2	2	1	2	2	1	2	1	1	3	1	2	1	3	3	2	1	2	2	1	3
19	2	1	2	2	1	1	2	2	1	2	1	1	2	1	3	3	3	1	2	2	1	2	3
20	2	1	2	2	1	1	2	2	1	2	1	2	3	2	1	1	1	2	3	3	2	3	1
21	2	1	2	2	1	1	2	2	1	2	1	3	1	3	2	2	2	3	1	1	3	1	2
22	2	1	2	1	2	2	2	1	1	1	2	1	2	2	3	3	1	2	1	1	3	3	2
23	2	1	2	1	2	2	2	1	1	1	2	2	3	3	1	1	2	3	2	2	1	1	3
24	2	1	2	1	2	2	2	1	1	1	2	3	1	1	2	2	3	1	3	3	2	2	1
25	2	1	1	2	2	2	1	2	2	1	1	1	3	2	1	2	3	3	1	3	1	2	2
26	2	1	1	2	2	2	1	2	2	1	1	2	1	3	2	3	1	1	2	1	2	3	3
27	2	1	1	2	2	2	1	2	2	1	1	3	2	1	3	1	2	2	3	2	3	1	1
28	2	2	2	1	1	1	1	2	2	1	2	1	3	2	2	2	1	1	3	2	3	1	3
29	2	2	2	1	1	1	1	2	2	1	2	2	1	3	3	3	2	2	1	3	1	2	1
30	2	2	2	1	1	1	1	2	2	1	2	3	2	1	1	1	3	3	2	1	2	3	2
31	2	2	1	2	1	2	1	1	1	2	2	1	3	3	3	2	3	2	2	1	2	1	1
32	2	2	1	2	1	2	1	1	1	2	2	2	1	1	1	3	1	3	3	2	3	2	2
33	2	2	1	2	1	2	1	1	1	2	2	3	2	2	2	1	2	1	1	3	1	3	3
34	2	2	1	1	2	1	2	1	2	2	1	1	3	2	3	2	1	2	3	1	2	2	1
35	2	2	1	1	2	1	2	1	2	2	1	2	1	3	1	3	2	3	1	2	3	3	2
36	2	2	1	1	2	1	2	1	2	2	1	3	2	1	2	1	3	1	2	3	1	1	3

3.4.1 SIMULATION AND DATA ANALYSIS

Let x_{ij} be the offset voltage corresponding to the ith row of the control orthogonal array and the jth row of the noise orthogonal array. Then,

$$r_i = (1/36) \sum_{j=1}^{36} x_{ij}^2$$

is the mean square offset voltage corresponding to the ith row of the control orthogonal array. Note that, here the mean square offset voltage is obtained using a deterministic simulation guided by the noise orthogonal array. Random number simulation is not used. However, the 36 values of x_{ij} that go in the computation of r_i represent a sample from the possible $2^{10} \times 3^{11}$ combinations of the noise parameter levels. Thus, we view r_i as a random variable in conducting ANOVA.

In the decibel scale, the mean square offset voltage is given by $\eta_i = -10 \log 10 r_i$. We have 36 values of η_i for the 36 rows of the control orthogonal array.

The analysis of variance of the η_i values is given in Table 3.5. We notice that only RPEM, CPCS and OCS have significant effect at 5% level on the values of η. The graphs in Figure 3.2 display these effects pictorially. Note that the factor level combination $B_1 D_1 E_1$ has a higher η value and hence a lower mean square offset voltage. In fact using the linear model with main effects we can predict the root mean square offset voltage to reduce from the starting value of 34.1 mV to 20.2 mV. A simulation was performed using the noise orthogonal array to estimate the root mean square offset voltage under the predicted optimum conditions and a value of 16.2 mV was obtained which is quite close to the predicted value of 20.2 mV.

TABLE 3.5. ANALYSIS OF VARIANCE TABLE

Factor	d.f.	Sum of Squares	F
A. RFM	2	9.9	0.53
B. RPEM	2	463.7	25.03*
C. RNEM	2	29.9	1.61
D. CPCS	2	111.1	6.00*
E. OCS	2	87.5	4.72*
Error	25	231.6	
Total	35	933.7	

Factors significant at 5% level are indicated by *

Thus we obtained a 40% reduction in the root mean square offset voltage which is a significant improvement in quality by simply shifting the mean levels of three of the control parameters. Note that this change in design does not result into any increase in the cost of the product.

Designed experiments and analysis of variance can be used to further explore the design space around the optimum point. Also it can be used to determine which tolerances should be tightened to further reduce the mean square offset voltage while minimizing the associated cost increase.

In summary here we have used orthogonal arrays for dual purposes—first to estimate the effect of the noise factors and second to study the multidimensional space spanned by the control factors.

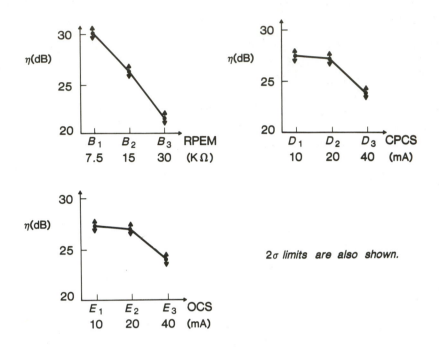

2σ limits are also shown.

Figure 3.2. Plots of Significant Factor Effects

3.5 LINEAR GRAPHS

Industrial experiments come in a great variety—different number of factors, different combinations of numbers of factor levels, different requirements on the estimation of interactions and different restrictions on the way experiments can be run. To design these experiments efficiently, we need a set of tables and simple procedures to construct orthogonal designs with minimum effort. Orthogonal arrays and their linear graphs developed by Taguchi[10] satisfy this need very well. To illustrate the utility of the linear graphs, consider an example of cutting boards with a router bit from panels. The objective of the experiment is to select the best router bit from among four possible bits and the optimum cutting conditions. The factors associated with this study and their numbers of levels are: A) Suction, 2 levels, B) X - Y feed, 2 levels, C) Type of bit, 4 levels, D) Spindle position, 4 levels, E) Type of suction foot, 2 levels, F) Stacking height, 2 levels, G) Depth of slot, 2 levels and H) Speed, 2 levels. The experimental design has the following requirements:

(i)   There are four spindles on a machine. All spindles are restricted to run at same speed and X - Y feed. Also, the experiment should be so set up that the effect of the spindle position is cancelled out in estimating the factorial effects of the other factors.

(ii)   It is difficult to change the suction. So it is preferable to make fewest changes in its level.

(iii) We want to study the B x F interaction.

A complete factorial design will need $2^7 \times 4^2 = 2048$ experiments. An orthogonally balanced fractional factorial experiment to satisfy the above restrictions can be obtained with the help of linear graphs as follows. Figure 3.3 gives the translation of the above restrictions into a linear graph. Consider the orthogonal array L_{16} which has 15 two-level columns (Taguchi and Wu[8]) which is given in Table 3.6. A four level column can be created in a two level orthogonal table by combining two columns and keeping their interaction column empty. This is indicated by two dots connected by a line. So for each of the factors C and D we need two dots connected by a line. We need the interaction B x F, so a line is drawn connecting these factors. In Figure 3.3, note that some factors are indicated by heavy dots while the others are indicated by hollow dots. Hollow dots mean that we want minimum changes in those factors, whereas a heavy dot means we want or can accommodate many changes in that factor. Factors A, B, and H should have hollow dots because they cannot be changed from spindle to spindle in a given machine run. Factor D should have heavy dots because we want all four spindles represented in a given machine run. There are no special restrictions on the remaining factors, they could be heavy or hollow dots.

Having prepared the linear graph of Figure 3.3, it is a rather easy matter to match it with one of the several linear graphs of L_{16} (Taguchi and Wu[8]) to come up with the column assignment indicated on the same figure. The resulting design is given in Table 3.7. Notice that experiments 1 through 4 have a common speed and X - Y feed; and all four spindles are represented in these four experiments. Thus experiments 1 through 4 form a machine run. Similarly experiments 5 through 8, 9 through 12 and 13 through 16 form convenient machine runs.

Taguchi[10] has tabulated many orthogonal arrays and their linear graphs. These arrays cover various combinations of 2, 3, 4 and 5 level columns. With linear graphs, these combinations can be extended to even more possibilities. Recently R. N. Kackar[11] has constructed some more orthogonal arrays. An excellent reference for the methods of constructing orthogonal arrays is Raghava Rao.[12]

TABLE 3.6. ORTHOGONAL ARRAY L_{16}

Expt. No.	Factor No.														
	1	2	3	4	5	6	7	8	9	10	11	12	13	14	15
1	1	1	1	1	1	1	1	1	1	1	1	1	1	1	1
2	1	1	1	1	1	1	1	2	2	2	2	2	2	2	2
3	1	1	1	2	2	2	2	1	1	1	1	2	2	2	2
4	1	1	1	2	2	2	2	2	2	2	2	1	1	1	1
5	1	2	2	1	1	2	2	1	1	2	2	1	1	2	2
6	1	2	2	1	1	2	2	2	2	1	1	2	2	1	1
7	1	2	2	2	2	1	1	1	1	2	2	2	2	1	1
8	1	2	2	2	2	1	1	2	2	1	1	1	1	2	2
9	2	1	2	1	2	1	2	1	2	1	2	1	2	1	2
10	2	1	2	1	2	1	2	2	1	2	1	2	1	2	1
11	2	1	2	2	1	2	1	1	2	1	2	2	1	2	1
12	2	1	2	2	1	2	1	2	1	2	1	1	2	1	2
13	2	2	1	1	2	2	1	1	2	2	1	1	2	2	1
14	2	2	1	1	2	2	1	2	1	1	2	2	1	1	2
15	2	2	1	2	1	1	2	1	2	2	1	2	1	1	2
16	2	2	1	2	1	1	2	2	1	1	2	1	2	2	1

TABLE 3.7. DESIGN MATRIX FOR THE ROUTING EXPERIMENT

Expt. No	Factor								
	A	B	H	C	D	E	F	G	BxF
1	1	1	1	1	1	1	1	1	1
2	1	1	1	2	2	1	2	2	2
3	1	1	1	4	4	2	2	1	2
4	1	1	1	3	3	2	1	2	1
5	1	2	2	1	4	1	2	2	1
6	1	2	2	2	3	1	1	1	2
7	1	2	2	4	1	2	1	2	2
8	1	2	2	3	2	2	2	1	1
9	2	1	2	1	3	2	2	1	2
10	2	1	2	2	4	2	1	2	1
11	2	1	2	4	2	1	1	1	1
12	2	1	2	3	1	1	2	2	2
13	2	2	1	1	2	2	1	2	2
14	2	2	1	2	1	2	2	1	1
15	2	2	1	4	3	1	2	2	1
16	2	2	1	3	4	1	1	1	2

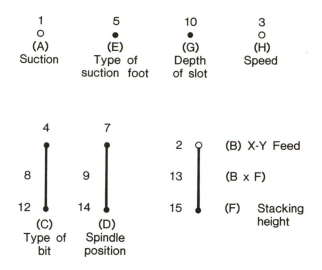

Figure 3.3. Linear Graph for Router Bit Experiment. The letters in the graph indicate factors; and the numbers indicate the corresponding columns of the orthogonal array L_{16}. To form a four level column for factor C, we combine columns 4 and 12 and leave their interaction columns 8 empty. A four level column for factor D is formed similarly. As indicated in the graph, interaction B x F is estimated from column 13.

3.6 CONCLUDING REMARKS

Through the preceding examples we have seen that design of experiments can contribute to a substantial increase in quality and productivity. In the case of the integrated circuit fabrication example, just 18 carefully chosen experiments gave us a 50% reduction in the process standard deviation, 67% reduction in the fatal defects due to window not open and at the same time a 50% reduction in the processing time. In the circuit design application, a modest simulation with $36^2 = 1296$ circuit analysis enabled us to reduce the root mean square offset voltage by 40%. And this improvement did not require any tighter component tolerances. So from the cost point of view, this is quality improvement without any cost increase.

The experimental design and ANOVA methods were originally developed by R. A. Fisher in the 1920's in conjunction with agricultural experiments. The theory and application of experimental design and response surface methods have been advanced by many researchers and today we have many excellent textbooks on this subject, e.g., Box, Hunter and Hunter,[13] Box and Draper,[14] Raghava Rao,[12] Kempthorne,[15] Hicks,[16] etc. The use of orthogonal arrays for design of experiments was first proposed by Rao.[17] Useful orthogonal designs have been developed by many researchers including Plackett and Burman[18] and Addelman[19] and Taguchi.[10]

In Japan the awareness and the use of the design of experiment methods is far more extensive when compared to that in the United States. For example, Nippon Denso, a well-known supplier of electrical parts all over the world used experiments based on orthogonal arrays 2700 times in 1976 (Taguchi and Wu[8]). Thousands of engineers are trained each year through company sponsored courses and courses sponsored by the Japanese Standards Association and the Japanese Union of Scientists and Engineers. Some accounts credit a major share of Japan's growth in quality and productivity over the last three decades to the extensive use of statistically design experiments.

This approach to design and analysis of experiments using orthogonal arrays and linear graphs which is widely used in Japan has the following desired features:

i)   It is simple to design complex experiments involving many factors at different number of levels. Analysis of the resulting data is also standardized and simple.

ii) Importance is given to both the mean and the variance of the response.

iii) A large number of factors are studied simultaneously which makes it an efficient experimentation technique. Only those interactions which are suspected to be large are included in the design.

I propose that we make design of experiments as a required part of engineering curriculum and also teach to practicing engineers. This will provide big improvements in efficiency, quality and productivity.

ACKNOWLEDGMENTS

I am grateful to Professor Genichi Taguchi for introducing to us his methods of quality control and to Roshan Chaddha for his invaluable support and encouragement throughout this project. I also thank my colleagues R. N. Kackar, D. V. Speeney, M. J. Grieco, G. M. Blaine and Dave Chrisman who made great contributions to the applications discussed in this paper.

REFERENCES

1. Taguchi, G. n.d. Seminar Given at AT&T Bell Laboratories (1979. *The Asahi*. April 17)

2. Duncan, A. J. 1974. *Quality Control and Industrial Statistics.* Richard D. Irwin, Inc.

3. Taguchi, G. 1978. "Off-line and On-line Quality Control Systems." *International Conference on Quality Control.* Tokyo, Japan.

4. Kackar, R. N., and M. S. Phadke. n.d. *An Introduction to Off-line and On-line Quality Control Methods.* Unpublished work.

5. Phadke, M. S., R. N. Kackar, D. V. Speeney, and M. J. Grieco. 1983. "Off-line Quality Control is Integrated Circuit Fabrication Using Design of Experiments." *The Bell System Technical Journal* vol. 62, no. 5 (May-June):1273-1309.

> *(Quality Control, Robust Design, and*
> *the Taguchi Method; Article 6)*

6. Phadke M. S., and G. M. Blaine n.d. *Quality Engineering in Circuit Design.* Forthcoming.

7. Box, G. E. P. n.d. "Evolutionary Operations: A Method for Increasing Industrial Productivity." *Applied Statistics* **6**:81-101.

8. Taguchi, G., and Y. Wu. 1980. *Introduction to Off-line Quality Control Systems.* Tokyo: Central Japan Quality Control Association.

9. Taguchi, G. 1974. "A New Statistical Analysis Method for Clinical Data, the Accumulation Analysis, in Contrast with the Chisquare Test." *Seishain-igaku* **29**:806-813.

10. Taguchi, G. 1976, 1977. *Experimental Designs.* 3d ed. Vol. 1 and 2. Tokyo: Maruzen Publishing Company. (Japanese)

11. Kackar, R. N. 1982. *Some Orthogonal Arrays for Screening Designs.* AT&T Bell Laboratories Technical Memorandum Unpublished work.

12. Rao, D. Raghava. 1971. *Construction and Combinatorial Problems in Design of Experiments.* New York: John Wiley & Sons.

13. Box, G. E. P., W. G. Hunter, and J. S. Hunter. 1978. *Statistics for Experimenters—An Introduction to Design, Data Analysis and Model Building.* New York: John Wiley & Sons.

14. Box, G. E. P., and N. R. Draper. 1969. *Evolutionary Operations.* New York: John Wiley and Sons.

15. Kempthorne, O. 1979. *The Design and Analysis of Experiments.* New York: Robert E. Krieger Publishing Co.

16. Hicks, C. R. 1973. *Fundamental Concepts in the Design of Experiments.* New York: Holt, Rinehart and Winston.

17. Rao, C. R. 1947. "Factorial Experiments Derivable from Combinatorial Arrangements of Arrays." *Journal of Royal Statistical Society, Ser. B* **9**:128-129.

18. Plackett R. L., and J. P. Burman. n.d. "The Design of Optimal Multifactorial Experiments." *Biometric* **33**:305-325.

19. Addelman, S. 1962. "Orthogonal Main-Effect Plans for Asymmetrical Factorial Experiments." *Technometrics* **4**:21-46.

ADDITIONAL REFERENCES

Cochran, W. G., and G. M. Cox. 1957. *Experimental Design*. New York: John Wiley & Sons.

Davis, O. L. *The Design and Analysis of Industrial Experiments*. New York: Imperial Chemical Industries Limited, Longman Group Limited.

Diamond, W. J. 1981. *Practical Experiment Design for Engineers and Scientists*. Lifetime Learning Publications.

4

OFF-LINE QUALITY CONTROL, PARAMETER DESIGN, AND THE TAGUCHI METHOD

Raghu N. Kackar

4.1 INTRODUCTION

A Japanese ceramic tile manufacturer knew in 1953 that is more costly to control causes of manufacturing variations than to make a process insensitive to these variations. The Ina Tile Company knew that an uneven temperature distribution in the kiln caused variation in the size of the tiles. Since uneven temperature distribution was an assignable cause of variation, a process quality control approach would have increased manufacturing cost. The company wanted to reduce the size variation without increasing cost. Therefore, instead of controlling temperature distribution they tried to find a tile formulation that reduced the effect of uneven temperature distribution on the uniformity of tiles. Through a designed experiment, the Ina Tile Company found a cost-effective method for reducing tile size variation caused by uneven temperature distribution in the kiln. The company found that increasing the content of lime in the tile formulation from 1% to 5% reduced the tile size variation by a factor of ten. This discovery was a breakthrough for the ceramic tile industry.

A technique such as this that reduces variation by reducing the sensitivity of an engineering design to the sources of variation rather than by controlling these sources is called *parameter design*. This example, which appeared in a Japanese book, "Frontier Stories in Industry" published by Diamond Sha Publishing Company in Japan, illustrates that parameter design is a cost-effective technique for improving manufacturing processes.

An even more important use of parameter design is for improving a product's field performance. A product's field performance is affected by environmental variables, product deterioration, and manufacturing imperfections. Parameter design can be used to make a product design robust against these sources of variation and hence improve field performance.

Variation in a product's performance during the product's life span is an important aspect of product quality. Off-line quality control methods reduce performance variation and hence the product's lifetime cost. A quantitative measure of performance variation is the expected value of monetary losses during the product's life span due to this variation. Parameter design experiments identify settings of product design characteristics that minimize the estimated expected loss. A performance statistic is a criterion for comparing different settings of product design characteristics that takes advantage of the prior engineering knowledge about the product and the loss function. Taguchi recommends the use of orthogonal arrays for planning parameter design experiments. Parameter design experiments can also be used to identify settings of the process variables that minimize the effect of manufacturing variations on the process's performance.

4.2 PERFORMANCE VARIATION

The American Society for Quality Control,[1] defines *quality* as "the totality of features and characteristics of a product or a service that bear on its ability to satisfy [a user's] given needs." *Performance characteristics* are the final characteristics of a product that determine the product's performance in satisfying a user's needs. The sharpness of the picture on a TV set is an example of a performance characteristic. Most products have many performance characteristics of interest to the user. In order to determine the degree of satisfaction with a performance characteristic, the ideal state of the performance characteristic from the customer's viewpoint must be known. This ideal state is called the *target value*.

All target specifications of performance characteristics should be stated in terms of nominal levels and tolerances around nominal levels. It is a widespread practice in industry to state target values of performance characteristics in terms of interval specifications only. This practice erroneously conveys the idea that a user remains equally satisfied for all values of the performance characteristic in the specification interval and then suddenly becomes completely dissatisfied the moment the performance value slips out of the specification interval. The target value should be defined as the ideal level of the performance characteristic, and it may or may not be the midpoint of the tolerance interval. For example, if "percentage impurity" is the performance characteristic and $(0,a)$ is the tolerance interval for some a, the target value is zero.

The degree of *performance variation*—the amount by which a manufactured product's performance deviates from its target value during the product's life span under different operating conditions and across different units of the product—is an importance aspect of product quality. The smaller the performance variation about the target value, the better the quality. For example, a TV set whose picture quality varies with weather conditions has poor quality. How does a customer feel if his automobile, with an intended life span of 100,000 miles, works for only 30,000 miles before falling apart? A good quality product performs at the target level consistently throughout the product's life span and under all different operating conditions. The degree of customer satisfaction is inversely proportional to the degree of performance variation. Juran[2] defines several aspects of product quality: quality of design, quality of conformance, and reliability. A product's performance variation is determined by all three of these important quality aspects.

The primary causes of a product's performance variation are: environmental variables, product deterioration, and manufacturing imperfections. Fluctuations in environmental variables such as temperature, humidity, and electrical power supply can cause variation in product performance. Product deterioration with age is an important cause of performance variation. Common examples of product deterioration are: increase in the electrical resistance of a resistor, loss of resilience of springs, and wearing out of moving parts in a motor due to friction. Further, it is inevitable in a manufacturing process to have some variation among different units of a product.

Although all stages of the product development cycle affect the quality and cost of a manufactured product, it is the designs of both the product and the manufacturing process that play crucial roles in determining the degree of performance variation and the manufacturing cost. A product's development cycle can be partitioned into three separate but overlapping stages: product design, process design, and manufacturing. At the product design stage engineers develop complete product design specifications including the specification of materials, components, configuration, and features. Next, process engineers design a manufacturing process. It may involve creation of a new process or modification of an existing process to produce the new product. The manufacturing department then uses the manufacturing process to produce many units of the product. As indicated in Table 4.1 (condensed from Taguchi,[3] and Kackar and Phadke[4]) countermeasures against performance variation caused by environmental variables and product deterioration can be built into the product only at the product design stage.

Manufacturing engineers are well aware that it is expensive to control a manufacturing process that is sensitive to manufacturing variations. Also, the cost of detection and correction of manufacturing imperfections increases rapidly as the product moves along a manufacturing line. Usually it is cheapest to correct manufacturing imperfections in a product immediately after these imperfections are formed. Because of increasing complexity of modern industrial and consumer products, product and process designs play an increasingly larger role in determining both the quality and the cost of a manufactured product.

TABLE 4.1. PRODUCT DEVELOPMENT STAGES AT WHICH COUNTER-
MEASURES AGAINST VARIOUS SOURCES OF VARIATION
CAN BE BUILT INTO THE PRODUCT

Product Development Stages	Sources of Variation		
	Environmental Variables	Product Deterioration	Manufacturing Variations
Product design	O	O	O
Process design	X	X	O
Manufacturing	X	X	O

O–Countermeasures possible.
X–Countermeasures impossible.

Yet, traditional quality control methods such as cause and effect diagrams, process capability studies, process quality control, control charts, and empirical Bayes charts (see Hoadley[5]) concentrate almost exclusively on manufacturing. These quality control activities at the manufacturing stage, which are conducted to keep the manufacturing process in statistical control and to reduce manufacturing imperfections in the product, are *on-line quality control methods*.

4.3 OFF-LINE QUALITY CONTROL

Off-line quality control methods are quality and cost control activities conducted at the product and the process design stages in the product development cycle. The overall aim of off-line quality control activities is to improve product manufacturability and reliability, and to reduce product development and lifetime costs. Leading manufacturers of high quality products use off-line quality control methods such as design reviews, sensitivity analyses, prototype tests, accelerated life tests, and reliability studies. However, these methods are neither as thoroughly developed nor as widely applied as on-line quality control methods. Industry needs well researched off-line quality control methods. In particular, industry needs scientific methods to identify product and process design specifications that reduce both the degree of performance variation and the manufacturing cost.

As with performance characteristics, all specifications of product and process design characteristics should be stated in terms of ideal nominal values and tolerances around these nominal values. It is a widespread practice in industry to state these specifications in terms of tolerance intervals only. This practice can mislead a manufacturer into producing products whose characteristics are barely inside the tolerance intervals. Such products have poor quality and reliability. Even if all the characteristics of a product are within their tolerance intervals, the product may not perform satisfactorily because of interactions among the product characteristics.

For example, if the size of a car door is near its lower tolerance limit and the size of the door frame is near its upper tolerance limit, the door may not close properly. A product performs best when all the characteristics of the product are at their ideal values. Further, the knowledge of ideal values of product and process design characteristics and the knowledge of loss due to deviation from these ideal values of product and process design characteristics and the knowledge of loss due to deviation from these ideal values encourages continuous quality improvement.

Taguchi[3] has outlined a three-step approach for assigning nominal values and tolerances to product and process design characteristics: *system design, parameter design,* and *tolerance design.* System design is the process of applying scientific and engineering knowledge to produce a basic functional prototype design. The prototype model defines the initial settings of product or process design characteristics.

Parameter design is an investigation conducted to identify settings that minimize (or at least reduce) the performance variation. A product or process can perform its intended function at many settings of its design characteristics. However, variation in the performance characteristic may change with different settings. This variation increases both product manufacturing and lifetime costs. The term parameter design comes from an engineering tradition of referring to product characteristics as product parameters. An exercise to identify optimal parameter settings is therefore called parameter design.

Tolerance design is a method for determining tolerances that minimize the sum of product manufacturing and lifetime costs. The final step in specifying product and process designs is to determine tolerances around the nominal settings identified by parameter design. It is still a common practice in industry to assign tolerances by convention rather than scientifically. Tolerances that are too narrow increase manufacturing cost, and tolerances that are too wide increase performance variation and, hence, a product's lifetime cost.

4.4 EXPECTED LOSS

Identifying optimal parameter settings requires specifying a criterion that is to be optimized. One such criterion is the *expected loss*—the expected value of monetary losses an arbitrary user of the product is likely to suffer at an arbitrary time during the product's life span due to performance variation. A measure of the degree of the performance variation is the expected loss. The concept of expected loss makes the problem of reducing performance variation concrete. It also provides a common basis for comparing variations in different performance characteristics.

Let Y be a value of the performance characteristic of interest and suppose the target value of Y is τ. The value Y can deviate from τ both during the product's life span and across different units of the product. In statistical terms Y is a random variable with some probability distribution (the parameters of the probability distribution of Y are the nominal settings of product design characteristics). Variations in Y cause losses to the product's user. Let $l(Y)$ represent the loss in terms of dollars suffered by an arbitrary user of the product at an arbitrary time during the product's life span due to deviation of Y from τ. Usually it is difficult to determine the actual form of $l(Y)$, but for many performance characteristics, a quadratic approximation to $l(Y)$ adequately represents economic losses due to performance variation. (Use of quadratic loss is not new. Indeed, quadratic loss is the basis of least square theory founded by Gauss in 1809.) The simplest quadratic loss function is

$$l(Y) = K(Y-\tau)^2, \tag{1}$$

where K is some constant (see Figure 4.1).

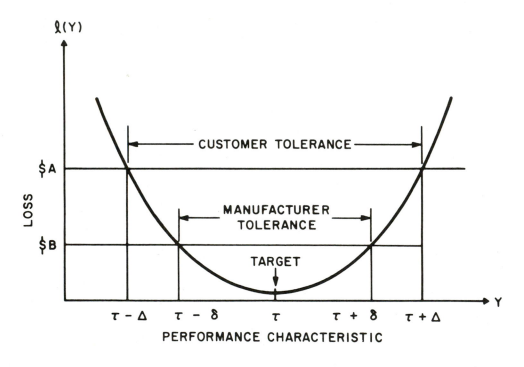

Figure 4.1. Loss Function

The unknown constant K can be determined if $l(Y)$ is known for any value of Y. Suppose ($\tau-\Delta$, $\tau+\Delta$) is the customer's tolerance interval. If a product performs unsatisfactorily when Y slips out of this interval, and the cost to the customer of repairing or discarding the product is A dollars, then from equation (1)

$$A = K\Delta^2 \text{ and } K = A/\Delta^2.$$

The manufacturer's tolerance interval $(\tau-\delta, \tau+\delta)$ can also be obtained from the loss function (1). Suppose before a product is shipped the cost to the manufacturer of repairing an item that exceeds the customer's tolerance limits is B dollars. Then

$$B = (A/\Delta^2)(Y-\tau)^2$$

$$Y = \tau \pm (B/A)^{\frac{1}{2}}\Delta$$

and

$$\delta = (B/A)^{\frac{1}{2}}\Delta.$$

Because B is usually much smaller than A, the manufacturer's tolerance interval will be narrower than the customer's tolerance interval (see Figure 4.1).

By taking the expectation of equation (1) with respect to the distribution of Y we get the expected loss

$$L = E[l(Y)] = KE[(Y-\tau)^2]. \tag{2}$$

The expectation in equation (2) is defined with respect to the distribution of Y, both during the product's life span and across different users of the units of the product.

In some situations, the loss function $l(Y)$ is not symmetric around τ. This would be the case, for example, if the performance characteristic Y has a nonnegative distribution and the target value of Y is zero. An asymmetric quadratic loss function has the form

$$l(Y) = \begin{cases} K_1(Y-\tau)^2 & \text{if } Y \leq \tau \\ \\ K_2(Y-\tau)^2 & \text{if } Y > \tau. \end{cases} \tag{3}$$

The unknown constants K_1 and K_2 can be determined if $l(Y)$ is known for a value of Y below τ and for a value of Y above τ. Suppose $(\tau-\Delta_1, \tau+\Delta_2)$ is the customer's tolerance interval and the cost to the customer of repair is A_1 dollars when Y falls below $\tau-\Delta_1$ and A_2 dollars when Y rises above $\tau+\Delta_2$. Then $K_1 = A_1/\Delta^2_1$ and $K_2 = A_2/\Delta^2_2$. Taking the expectation of equation (3) results in the expected loss

$$L = E[l(Y)] \tag{4}$$

$$= K_1 P_1 E[(Y-\tau)^2 \mid Y\leq\tau] + K_2 P_2 E[(Y-\tau)^2 \mid Y>\tau]$$

where $P_1 = \Pr[Y \leq \tau]$ and $P_2 = 1-P_1$. If Y is nonnegative and $\tau = 0$; then $K_1 = 0$, $K_2 = K$ (say), the loss function (3) reduces to

$$l(Y) = KY^2 \qquad if\ Y \geq 0 \tag{5}$$

and the expected loss (4) reduces to

$$L = KE[Y^2]. \tag{6}$$

The expected losses (2), (4), and (6) can be estimated if a random sample from the distribution of Y can be obtained.

4.5 PARAMETER DESIGN FOR IMPROVING PRODUCT DESIGNS

Parameter design is an investigation conducted to identify settings of product design characteristics that minimize the expected loss. Most applications of parameter design require some kind of experimentation. These experiments can be either physical experiments or computer based simulation trials. Because of advances in computer-aided simulation tools, parameter design experiments for product design improvement are often done with a computer.

Parameter design is based on classifying the variables that affect a product's performance into two categories: *design parameters*[1] and *sources of noise*. Design parameters are the product design characteristics whose nominal settings can be specified by the product designer. A vector of the settings of design parameters defines a product design specification and vice versa. The actual values of design parameters in a manufactured product may deviate from the nominal settings. Sources of noise are all those variables that cause performance variation both during a product's life span and across different units of the product.

---

1. Glossary equivalent is *control parameter*.

4.5.1 NOISE CATEGORIES

Sources of noise can be classified into two categories: *external sources of noise* and *internal sources of noise*. External sources of noise are variables external to a product that affect the product's performance. Common examples are: variations in environmental variables such as temperature, humidity, dust, and vibrations; and human variations in operating the product. Internal sources of noise are the deviations of the actual characteristics of a manufactured product from the corresponding nominal settings. The primary internal sources of noise are: manufacturing imperfections and product deterioration.

Sometimes external sources of noise affect a product's performance indirectly by increasing the deviations of actual product characteristics from the nominal settings. For example, high temperature and humidity can make a product deteriorate faster. These deviations are internal sources of noise. Therefore, if a product's performance is insensitive to internal sources of noise, it will also be insensitive to those external sources of noise that affect performance by way of increasing the level of internal sources of noise.

Not all sources of noise can be included in a parameter design experiment because of physical limitations and because of lack of knowledge. *Noise factors*[2] are those sources of noise that can be included in a parameter design experiment. The noise factors should represent the sources of variation that affect a product's performance in the field. Parameter design is possible because the effects of both internal and external noise factors can change with the settings of the design parameters. As illustrated by the following examples from Taguchi and Wu,[6] it is possible to identify settings that reduce performance variation.

4.5.2 EXAMPLES

Example 1: The product is a type of caramel produced by a Japanese company in 1948. Caramel is a mixture of more than ten ingredients but is mostly sugar. The performance characteristic of interest was the plasticity of the caramel. The plasticity of the caramel produced by the company was very sensitive to environmental temperature as shown in Figure 4.2. Caramels whose plasticity were less sensitive to environmental temperature were considered to be of higher quality. A parameter design experiment was conducted by the company in order to develop a caramel formulation whose plasticity was less sensitive to environmental temperature. The improvement was possible because the effect of the

---

2. Glossary equivalent is *noise parameter*.

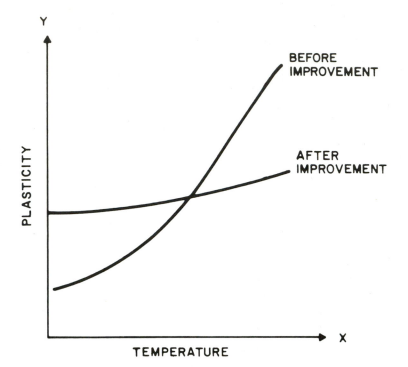

Figure 4.2. Effect of Temperature on the Plasticity of Caramel

external noise factor temperature on the plasticity of caramel changed with the caramel formulation.

Example 2: Consider an electrical circuit. Suppose the performance characteristic of interest is the output voltage of the circuit and its target value is y_0. Assume that the output voltage of the circuit is largely determined by the gain of a transistor X in the circuit and the circuit designer is at liberty to choose the nominal value of this transistor. Suppose also that the effect of transistor gain on the output voltage is nonlinear as shown in Figure 4.3.

In order to obtain an output voltage of y_0, the product designer can select the nominal value of the transistor gain to be x_0. If the actual transistor gain deviates from the nominal value x_0, the output voltage will deviate from y_0. The transistor gain can deviate from x_0 because of manufacturing imperfections in the transistor, because of deterioration during the circuit's life span, and because of environmental variables. The variation of the actual transistor gain around its nominal value

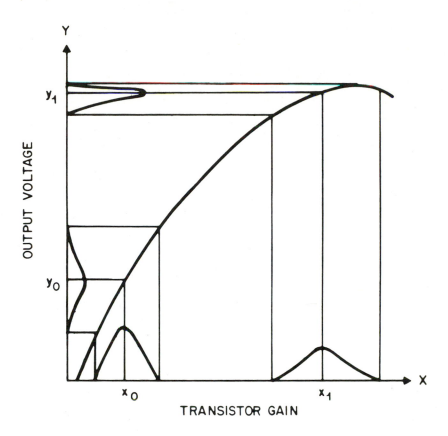

Figure 4.3. Effect of Transistor Gain on the Output Voltage

is an internal noise factor in the circuit. If the distribution of this internal noise factor is as shown in Figure 4.3, the output voltage will have a mean value equal to y_0, but with a large variance. On the other hand if the circuit designer selects the nominal value of the transistor gain to be x_1, the output voltage will have a much smaller variance. But the mean value of the output voltage y_1 associated with the nominal value x_1 is far from the target value y_0. Now suppose there is another component in the circuit that has a linear effect on the output voltage and the circuit designer is at liberty to choose the nominal value of this component. The product designer can then adjust the value of this component to move the mean value of output voltage from y_1 to the target value y_0. Adjustment of the mean value of the performance characteristic to its target value is usually a much easier engineering problem than the reduction of performance variation. From the output voltage viewpoint changing the nominal transistor gain from x_0 to x_1

improves the circuit design. Of course, this change would not necessarily improve the circuit design if it were accompanied by an increase in the variance of another performance characteristic of the circuit.

4.5.3 PARAMETER DESIGN EXPERIMENT

A parameter design experiment consists of two parts: a *design matrix*[3] and a *noise matrix*.[4] The columns of a design matrix represent the design parameters, entries in the columns represent test settings of the design parameters, and each row of the matrix (also called a test run) represents a product design. The columns of a noise matrix represent noise factors, and the rows of the matrix represent different combinations of the levels of the noise factors. A complete parameter design experiment consists of a combination of design and noise matrices as indicated in Figure 4.4. If the design matrix has m rows and the noise matrix has n rows, the total number of rows in the combined parameter design experiment is $m \times n$. For each of the m rows of the design matrix, the n rows of the noise matrix provide n or more repeat observations on the performance characteristic. The levels of the noise factors and the noise matrix are chosen so that these repeat observations are representative of the effects of all possible levels of the noise factors. The repeat observations on the performance characteristic from each test run in the design matrix are then used to compute a criterion called a performance statistic. The m values of the performance statistic associated with the m test runs in the design matrix are then used to predict settings of design parameters that minimize the expected loss.

4.6 PERFORMANCE STATISTICS

A *performance statistic* estimates the effect of noise factors on the performance characteristic. An efficient performance statistic takes advantage of the prior engineering knowledge about the product, the loss function, and the distribution of the performance characteristic. Let $\theta = (\theta_1, \theta_2, \cdots, \theta_k)$ represent design parameters and let $\omega = (\omega_1, \omega_2, \cdots, \omega_l)$ represent the noise factors included in the parameter design experiment. We assume that the performance characteristic Y is a function of θ and ω, that is, $Y = f(\theta, \omega)$. The design parameters θ are the parameters of the distribution of Y, and for a given θ the noise factors generate the distribution.

---

3. Glossary equivalent is *control array.*
4. Glossary equivalent is *noise array.*

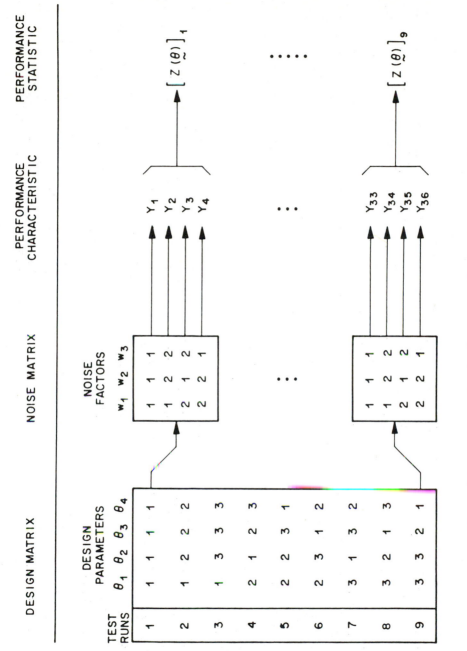

Figure 4.4. An Example of a Parameter Design Experiment Plan

Let

$$\eta\,(\,\theta\,)\ = E\,[Y]$$

and

$$\sigma^2\,(\theta) = E\,[\{Y - \eta(\theta)\}^2]$$

represent the mean and variance of Y, respectively. With this model expected losses are functions of θ.

A *performance measure* is a function of θ chosen so that maximization of the performance measure along with possible engineering adjustments minimizes the expected loss. The performance measure is then used as a criterion for comparing different settings of the design parameters. An efficient performance measure takes advantage of engineering knowledge about the product (to possibly simplify the measure) and the loss function. The expected loss itself is a performance measure, however, sometimes it is more complicated than necessary because it does not take advantage of engineering knowledge. Different engineering designs can lead to different performance measures. Taguchi has told us that he has defined more than 60 different signal to noise ratios (estimates of performance measures) for engineering applications of parameter design. While a performance measure is a function of θ, in general the function is not known. It must, therefore, be estimated; and it is this estimate that is used as the criterion to be optimized. We use the term *performance statistic* for a statistical estimate of a performance measure, whereas Taguchi[7] uses the term signal to noise ratio. (Table 4.2 shows the correspondence between the terminologies of Taguchi and this paper.)

TABLE 4.2. CORRESPONDENCE BETWEEN THE TERMINOLOGIES OF TAGUCHI AND THIS PAPER

This Paper	Taguchi
Performance variation	Functional variation
Performance statistic	Signal to noise ratio
Design parameters	Control factors
External noise factors	Outer noise factors
Internal noise factors	Inner noise factors
Design matrix	Inner array
Noise matrix	Outer array

When the performance characteristic Y is a continuous variable, the loss function $l(Y)$ usually takes on one of three forms depending on whether smaller is better, larger is better, or a specific target value is best. For the first two cases we will discuss the performance statistics recommended by Taguchi. For the third case we will show how two different engineering situations (one considered by Taguchi and one not) lead to difference performance statistics. Let

$$MSE(\theta) = E[(Y - r)^2]$$

$$B(\theta) = \eta\ (\theta) - r$$

and

$$\xi(\theta) = [\eta\ (\theta)]^2/\sigma^2(\theta)$$

represent the mean squared error, the bias, and the square of the inverse of the coefficient of variation of the distribution of Y.

4.6.1 THE SMALLER THE BETTER

The performance characteristic Y has a nonnegative distribution, the target value is $\tau = 0$, and the loss function $l(Y)$ increases as Y increases from zero. In this case the expected loss (6) is proportional to

$$MSE(\theta) = E[(Y - 0)^2] = E[Y^2]$$

and Taguchi[7] recommends using the performance measure

$$\zeta(\theta) = -10\ log\ MSE(\theta). \tag{7}$$

The larger the performance measure the smaller is the mean squared error. Let $y_1, y_2, \cdots, y_n$ approximate a random sample from the distribution of Y for a given θ. The performance measure (7) can then be estimated by the performance statistic

$$Z(\theta) = -10\ log\ (\textstyle\sum y_i^2/n). \tag{8}$$

The performance statistic (8) is the method of moments estimator of (7).

4.6.2 THE LARGER THE BETTER

The performance characteristic Y has a nonnegative distribution, the target value is infinity, and the loss function $l(Y)$ decreases as Y increases from zero. This is a particular application of the smaller the better case where we treat the reciprocal $1/Y$ as the performance characteristic. The target value of $1/Y$ is zero, the performance measure (7) reduces to

$$\zeta(\theta) = -10 \ log \ MSE(\theta)$$

where

$$MSE(\theta) = E[(1/Y)^2]$$

and the performance statistic (8) reduces to

$$Z(\theta) = -10 \ log \ (\textstyle\sum (1/y_i)^2/n).$$

4.6.3 A SPECIFIC TARGET VALUE IS BEST

The performance characteristic Y has a specific target value $\tau = \tau_0$, and the loss function $l(Y)$ increases as Y deviates from τ_0 in either direction. In this case the expected loss (2) is proportional to $MSE(\theta) = E[(Y - \tau_0)^2]$, and

$$MSE(\theta)' = \sigma^2(\theta) + [B(\theta)]^2 \tag{9}$$

$$= \sigma(\theta) + [\eta \ (\theta) - \tau_0]^2.$$

Variance Not Linked to Mean. In many engineering applications $\sigma(\theta)$ and $\eta(\theta)$ are functionally independent of each other, and the bias $B(\theta)$ can be reduced independently of the variance $\sigma^2(\theta)$. Often this can be accomplished simply by adjusting one or more design parameters called *adjustment parameters*. The adjustment parameters are special design parameters that can be easily adjusted to reduce a component (to be referred to as the adjustable component) of the total variation about the target value without affecting the remaining non-adjustable component of the total variation. In the present situation, $B(\theta)$ is the adjustable component of the mean squared error, $\sigma^2(\theta)$ is the non-adjustable component of the mean squared error, and the adjustment parameters are those design parameters that have a large effect on $\eta(\theta)$, and hence $B(\theta)$, but almost no effect on $\sigma^2(\theta)$. When such adjustment parameters are available, $B(\theta)$ can be reduced independently of $\sigma^2(\theta)$. Therefore, in this situation the real issue is to reduce $\sigma^2(\theta)$. So an efficient performance measure should be a monotonic function of $\sigma^2(\theta)$.

Thus

$$\zeta(\theta) = 10 \, log \, \sigma^2(\theta) \qquad (10)$$

can be used as a performance measure. The larger the performance measure the smaller is the variance $\sigma^2(\theta)$. Let $y_1, y_2, \cdots, y_n$ approximate a random sample from the distribution of Y for a given θ. Then

$$s^2 = \sum (y_i - \bar{y})^2 / (n-1),$$

where

$$\bar{y} = \sum y_i / n$$

is an (unbiased) estimator of $\sigma^2(\theta)$ and

$$Z(\theta) = -log(s^2) \qquad (11)$$

is a reasonable performance statistic. (This is not a Taguchi performance statistic.) Often the log transformation reduces curvatures and interactions in the s^2 response surface. In addition, log transformation often reduces the dependence of s^2 on $\bar{y}$. The minus sign in (11) is used by convention so that we always maximize performance statistics. Bartlett recommended the use of the[8] performance statistic (11) for testing the equality of several population variances.

Variance Linked to Mean. Sometimes $\sigma(\theta)$ increases linearly with $\eta(\theta)$, but the bias $B(\theta)$ can be reduced independently of the coefficient of variation $\sigma(\theta)/\eta(\theta)$. In this case $B(\theta)$ is the adjustable variation; the coefficient of variation, or equivalently $\xi(\theta)$, is the non-adjustable variation; and the adjustment parameters are those design parameters that have a large effect on $\eta(\theta)$, and hence $B(\theta)$, but almost no effect on $\xi(\theta)$. When such adjustment parameters are available, it is relatively easy to reduce $B(\theta)$, and the crux of parameter design is to maximize $\xi(\theta)$. If $\xi(\theta)$ remains large, or equivalently the coefficient of variation remains small, any subsequent change in the mean value $\eta(\theta)$ does not bring about a large change in the variance $\sigma^2(\theta)$. So an efficient performance measure should be a monotonic function of $\xi(\theta)$. Taguchi[7] recommends using the performance measure

$$\zeta(\theta) = 10 \, log \, [\xi(\theta)]. \qquad (12)$$

The larger the performance measure the smaller is the coefficient of variation $\sigma(\theta)/\eta(\theta)$. The quantity $n\bar{y}^2 - s^2$ is an (unbiased) estimator of $n[\eta(\theta)]^2$. Therefore, Taguchi[7] recommends the performance statistics

$$Z_1(\theta) = 10 \log(\bar{y}^2/s^2) \tag{13}$$

and

$$Z_2(\theta) = 10 \log[(n\bar{y}^2 - s^2)/(ns^2)]. \tag{14}$$

Since we can write (14) as

$$Z_2(\theta) = 10 \log[(\bar{y}^2/s^2) - (1/n)]$$

both performance statistics are equivalent. The mean squared error and the expected loss are minimized when performance measures (10) and (12) are maximized, and the bias ($\eta(\theta) - \tau_0$) is minimized by the use of adjustment parameters.

4.7 THE TAGUCHI METHOD

The Taguchi Method for identifying settings of design parameters that maximize a performance statistic is summarized below.

1. Identify initial and competing settings of the design parameters, and identify important noise factors and their ranges.

2. Construct the design and noise matrices, and plan the parameter design experiment.

3. Conduct the parameter design experiment and evaluate the performance statistic for each test run of the design matrix.

4. Use the values of the performance statistic to predict new settings of the design parameters.

5. Confirm that the new settings indeed improve the performance statistic.

Now we discuss these steps in more detail.

1. A prototype model of the product, obtained from the system design, defines the initial settings of design parameters $\theta = (\theta_1, \theta_2, \cdots, \theta_k)$. The product designer then identifies other possible values of θ. The set of all possible values of θ is called the parameter space and denoted by Θ.

   The product designer identifies the noise factors that cause most of the performance variation. These noise factors are included in the parameter design experiment. Next, the ranges of the noise factors within which product performance is desired to be insensitive are identified. The set of all possible levels of the noise factors is called the noise space.

For a given θ in the parameter space Θ, an experiment is conducted at different levels of the noise factors to obtain repeat values of the performance characteristic. These repeat values represent the effect of the noise factors on the performance characteristic and are used to compute the performance statistic. We seek a value of θ in Θ that maximizes the performance statistic. There can be many such values and any one of them will suffice, provided there is no cost differential.

2. A thorough search of the parameter space Θ is not cost-effective. Statistical design of experiment methods can be used for selecting an intelligent subset of Θ called the design matrix. The columns of the design matrix represent the design parameters and the rows represent different settings of θ. A performance statistic is evaluated for every θ in the design matrix. In statistical terms, the performance statistic is the response variable. The choice of a performance statistic may have some consequences on the criteria that are used for selecting the design matrix. The design matrix should be selected so that the computed values of the performance statistic provide good information in a minimum number of test runs about a θ in Θ that maximizes the performance measure.

Taguchi[7] recommends the use of orthogonal arrays (more precisely orthogonal arrays of strength two) for constructing design matrices. Orthogonal arrays are generalized Graeco-Latin squares. The design matrix in Figure 4.4 is a Graeco-Latin square written in an orthogonal array form. All common fractional factorial designs are orthogonal arrays. Rao[9] is responsible for the general theory of orthogonal arrays. Raghavarao[10] gives several methods for constructing orthogonal arrays, and Kackar[11] provides a catalog of important orthogonal arrays. Orthogonal arrays allow the design parameters to have different numbers of test settings. Taguchi recommends that three or more test settings should be chosen for each design parameter. Three or more test settings reveal nonlinearities in the main effects of design parameters. The test settings should be chosen wide apart so that the design matrix covers a wide region of the parameter space.

Orthogonal array experiments have the pairwise balancing property that every test setting of a design parameter occurs with every test setting of all other design parameters the same number of times. Thus any two columns of an orthogonal array form a two factor complete factorial design. This makes a comparison of different test settings of a design parameter valid over the ranges implied by the test settings of other design parameters. However, this comparison can be misleading in the presence of interactions among design parameters. Hunter[12] discusses this and other issues in more detail.

Orthogonal array experiments minimize the number of test runs while keeping the pairwise balancing property. If the number of test runs is not an overriding concern, either larger balanced experiments such as fractional factorial designs of resolution IV or V, or response surface designs (see Box, Hunter, and Hunter[13]) can be used for constructing the design matrix.

Taguchi[7] also recommends the use of orthogonal arrays for constructing the noise matrix. The noise matrix is a selective rather than a random subset of the noise space. If the test levels of noise factors are judiciously chosen, orthogonal arrays enable an appropriate coverage of the noise space. Taguchi's guiding principle for selecting the test levels of noise factors is as follows. Suppose the mean value and the standard deviation of the distribution of a noise factor w_i are approximately M_i and S_i respectively. If w_i is assumed to have a linear effect on the performance characteristic Y, then it should have two test levels: $(M_i - S_i)$ and $(M_i + S_i)$. If w_i is assumed to have a quadratic effect on Y, then it should have three test levels: $(M_i - \sqrt{(3/2)}\,S_i)$, M_i, and $(M_i + \sqrt{(3/2)}\,S_i)$. These choices of the test levels are apparently based on the assumption that noise factors have approximately symmetric distributions. The mean value and the standard deviation of the three numbers $(M_i - \sqrt{(3/2)}\,S_i)$, M_i, and $(M_i + \sqrt{(3/2)}\,S_i)$ are, respectively, M_i and S_i. The complete parameter design experiment is a combination of the design and noise matrices as indicated in Figure 4.4.

3. A parameter design experiment can be conducted in one of two ways: (1) through physical experiments with experimental prototypes of the product, or (2) through computer simulation trials when the function $Y=f(\theta, \omega)$ can be numerically evaluated. Often the function $Y=f(\theta, \omega)$ is too complicated for analytical solution and simulation methods are appropriate.

When physical experiments are used, it is either impossible or very expensive to make many experimental prototypes with different levels of manufacturing imperfections. Since the aim of parameter design is to identify settings of design parameters that allow for wide manufacturing imperfections, inferior grade materials and parts are used to make the prototypes. The performance of these prototypes is then evaluated in the presence of different levels of external noise factors. This is the essence of Taguchi and Wu's[6] statement, "nothing is more foolish than research using high priced raw materials or component parts."

When numerical evaluation of the function $Y=f(\theta, \omega)$ is possible, more informed analyses can be performed. In particular, product performance can be evaluated in the presence of both internal and external noise factors. In the case of computer runs, the function $Y=f(\theta, \omega)$ is evaluated for each θ in the design matrix and for each ω in the noise matrix. The numerical values of Y are in turn used to compute the performance statistic for each θ in the design matrix.

When numerical evaluation of functions relating product performance characteristics, design parameters and noise factors is possible, several other methods for product design optimization are available. For example, Brayton, Director, and Hachtel,[14] and Singhal and Pinel[15] describe several numerical algorithms for circuit design optimization. Although both Singhal and Pinel[15] and Taguchi tackle the same problem, Singhal and Pinel[15] do not use the design of experiment methods and optimize many performance characteristics simultaneously while Taguchi optimizes each performance characteristic separately by using design of experiment methods.

4. Computed values of the performance statistic are used to predict a θ in the parameter space that maximizes the performance statistic. The repeat observations on the performance characteristic used to compute the performance statistic are generated by the noise matrix. Although the noise matrix is a selective set, the repeat observations on Y for a given θ obtained by the use of a noise matrix are assumed to approximate a random sample from the distribution of Y given θ.

Since orthogonal arrays are generalized Graeco-Latin squares, the data analysis procedure is also a generalization of the procedure for Graeco-Latin square experiments (see, e.g., Scheffe,[16] Chapter 5). The statistical model underlying the use of orthogonal arrays for constructing the design matrix is an additive model, which can be written as

$$[Z(\theta)]_{st\cdots u} = \mu_0 + [\theta_1]_s + [\theta_2]_t + \cdots + [\theta_k]_u + r \tag{15}$$

where $[\theta_i]_j$ represents the effect of the jth test setting of the design parameter θ_i on the performance statistic $Z(\theta)$ for $i = 1,2,\cdots,k$ and $j = 1,2,3,\cdots$; and r represents the residual. The residuals r for different values of θ are assumed to have independent and identical distributions with zero mean and constant variance. The computed values of $Z(\theta)$ are used to estimate the effects $[\theta_i]_j$. The estimated values of $[\theta_i]_j$ are in turn used to identify the test setting of each design parameter that gives a higher value of $Z(\theta)$. These test settings define new settings of the design parameters θ. If the additive model (15) is a reasonable approximation to the true situation, the new settings will hopefully be better than the initial settings.

5. It is not possible to know whether a statistical model underlying the design and analysis of an experiment is true. Therefore, it is necessary to confirm by follow-up experiments that the new settings improve the performance statistic over its value at the initial settings. A successful confirmation experiment alleviates concerns about possibly improper assumptions underlying the model. If the increase in the performance statistic obtained is not large enough, another iteration of parameter design may be necessary.

71

4.8 PARAMETER DESIGN FOR IMPROVING PROCESS DESIGNS

As indicated by the Ina Tile example, parameter design can reduce manufacturing imperfections without increasing manufacturing cost. Parameter design experiments can identify operating standards of controllable process variables at which the effect of manufacturing variations on process performance is minimum. Intermediate in-process product characteristics determine actual characteristics of the final product. Therefore, these in-process characteristics are used to evaluate process performance. In the process design context, design parameters are the process variables whose operating standards can be chosen by the process designer, and manufacturing variations are the sources of noise. Common examples of manufacturing variations are: variations in incoming materials and supplies, inability of the manufacturing equipment to provide exactly the same environmental conditions (temperature, concentration, vapor pressure, and humidity) to all units in a production batch, process operator variations, drift in operating conditions, and electrical power fluctuations.

When conducting parameter design experiments in manufacturing process situations, frequently it is not cost-effective to explicitly include many noise factors in the experiment. However, experiments can usually be planned so that the variation in repeat values from a test run adequately reflects the effect of manufacturing variations. These repeated values are then used to compute and maximize performance statistics.

Statistically designed experiments have been applied to improve industrial processes for more than 50 years (see, e.g., Tippett,[17] Davies,[18] Daniel,[19] and Box, Hunter, and Hunter.[13] However, most applications optimize the mean value of a response variable. Parameter design experiments are aimed at reducing variability caused by manufacturing variations. In industrial processes, controlling variability is usually much harder than controlling the mean value, and variability is often a root cause of high manufacturing costs.

Lin and Kackar[20] showed how a 36 run orthogonal array design was used to improve a wave soldering process by studying 17 variables simultaneously. Pao, Phadke, and Sherrerd[21] show how a parameter design experiment was used to optimize the response time of a computer operating system. Phadke and others[22] illustrate how a parameter design experiment was used to improve the photolithographic process in integrated circuit fabrication. Prasad[23] and Taguchi and Wu[6] provide many examples of parameter design experiments.

4.9 CONCLUDING REMARKS

While working in the Bell Telephone Laboratories in the 1920's, Shewhart[24] laid down the foundation of statistical methods for quality control. Shewhart pointed out that variation is a fact of life. The evolution of quality control

methods has led to better and better countermeasures against undesirable variation. Taguchi's work is major contribution to this evolution.

Following Ishikawa,[25] we can partition the evolution of quality control
efforts into three generations: 1) inspection, 2) manufacturing process control, and
3) product and process design improvement (see Figure 4.5).

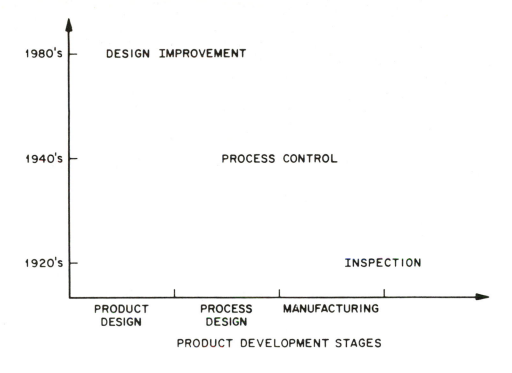

Figure 4.5. Quality Evolution

Inspection screens out nonconforming products before they reach the customer and
it provides feedback to the manufacturing department. Inspection does not
improve the built-in quality of the product. Manufacturing process controls reduce
manufacturing imperfections, but process controls cost money. Process design
improvement methods such as parameter design can reduce manufacturing imperfections without increasing cost. Manufacturing imperfections are only one aspect
of the quality of a product. An aim of product design improvement methods is to
make product performance insensitive to environmental variables, product
deterioration, and manufacturing imperfections. Improved product and process
designs reduce both manufacturing and product lifetime costs.

Statistical methods for inspection and manufacturing process controls have evolved over the last 60 years. However, development of statistical methods for product design improvement has just begun. Taguchi[7] has promoted the use of statistical design of experiment methods for product design improvement. He has given us an excellent starting point for further research in statistical methods for product design improvement as well as a stimulus for wider application of existing methods. Opportunities such as this to make a major contribution to industry do not come often.

ACKNOWLEDGMENTS

Suggestions by my colleagues Jeff Hooper, Ramon Leon, and Anne Shoemaker have clarified some of the issues. The following individuals also provided many helpful suggestions: Bob Easterling, Richard Freund, Gerry Hahn, Bruce Hoadley, Bob Kerwin, Bill Meeker, Peter Nelson, Vasant Prabhu, Steve Vardeman, and the referees.

[Discussions of this article and the author's response appear in *Journal of Quality Technology* **17** (Oct. 1985): 189-210.]

REFERENCES

1. ASQC (1983). *Glossary and Tables for Statistical Quality Control*. Milwaukee, Wis.: American Society for Quality Control.

2. Juran, J. M. 1979. *Quality Control Handbook*. 3d ed. New York: McGraw-Hill.

3. Taguchi, G. 1978. "Off-Line and On-Line Quality Control Systems." *Proceedings of International Conference on Quality Control*. Tokyo, Japan.

4. Kackar, R. N., and M. S. Phadke. n.d. *An Introduction to Off-line and On-line Quality Control Methods*. Unpublished work.

5. Hoadley, B. 1981. "The Quality Measurement Plan (QMP)." *The Bell System Technical Journal* **60**:215-273.

6. Taguchi, G., and Y. Wu. 1980. *Introduction to Off-line Quality Control*. Tokyo, Japan: Central Japan Quality Control Association. Available from American Supplier Institute, 32100 Detroit Industrial Expressway, Romulus, Michigan 48174.

7. Taguchi, G. 1976, 1977. *Experimental Designs*. 3d ed. 2 vols. Tokyo: Maruzen Publishing Company. (Japanese)

8. Bartlett, M. S. 1937. "Properties of Sufficiency and Statistical Tests." *Proceedings of the Royal Society, London, Series A* **160**:268-282.

9. Rao, C. R. 1947. "Factorial Experiments Derivable from Combinatorial Arrangements of Arrays." *Journal of the Royal Statistical Society, Supplement* **9**:128-139.

10. Raghavarao, D. 1971. *Constructions and Combinatorial Problems in Design of Experiments*. New York: John Wiley and Sons, Inc.

11. Kackar, R. N. 1982. *Some Orthogonal Arrays for Screening Designs*. AT&T Bell Laboratories Technical Memorandum. Unpublished work.

12. Hunter, J. S. 1985. "Statistical Design Applied to Product Design." *Journal of Quality Technology* **17**:210-221.

13. Box, G. E. P., W. G. Hunter, and J. S. Hunter. 1978. *Statistics for Experimenters*. New York: John Wiley and Sons, Inc.

14. Brayton, R. K., S. W. Director, and G. D. Hachtel. 1980. "Yield Maximization and Worst-Case Design with Arbitrary Statistical Distributions." *IEEE Transactions on Circuits and Systems* **CAS-27**:756-764.

15. Singhal, K., and J. F. Pinel. 1981. "Statistical Design Centering and Tolerancing Using Parametric Sampling." *IEEE Transactions on Circuits and Systems* **CAS-28**:692-702.

16. Scheffe, H. 1959. *Analysis of Variance*. New York: John Wiley and Sons, Inc.

17. Tippett L. C. H. 1934. *Applications of Statistical Methods to the Control of Quality in Industrial Production*. England: Manchester Statistical Society.

18. Davies, O. L. 1978. *The Design and Analysis of Industrial Experiments*. New York: Longman, Inc.

19. Daniel, C. 1976. *Applications of Statistics to Industrial Experimentation*. New York: John Wiley and Sons, Inc.

20. Lin, K. M., and R. N. Kackar. 1985. "Wave Soldering Process Optimization by Orthogonal Array Design Method." *Electronic Packaging and Production*.

21. Pao, T. W., M. S. Phadke, and C. S. Sherrerd. 1985. "Computer Response Time Optimization Using Orthogonal Array Experiments." *Proceedings of ICC*. IEEE Communications Society.

22. Phadke, M. S., R. N. Kackar, D. V. Speeney, and M. J. Grieco. 1983. "Off-Line Quality Control for Integrated Circuit Fabrication Using Experimental Design." *The Bell System Technical Journal* **62**:1273-1309.

23. Prasad, C. R. 1982. *Statistical Quality Control and Operational Research: 160 Case Studies in Indian Industries*. Calcutta, India: Indian Statistical Institute.

24. Shewhart, W. A. 1931. *Economic Control of Quality of a Manufactured Product.* New York, N.Y.: D. Van Nostrand Company. 1981. Reprint. American Society for Quality Control.

25. Ishikawa, K. 1984. "Quality and Standardization: Progress for Economic Success." *Quality Progress* **1**:16-20.

5

QUALITY ENGINEERING THROUGH DESIGN OPTIMIZATION

G. Taguchi and M. S. Phadke

5.1 INTRODUCTION

A survey done by J. M. Juran[1] showed that the Japanese yields of LSI chips are higher than those of the American and European manufacturers by a factor of two to three. Another study made by K. B. Clark[2] showed that the design and development cycle for a specific product was about 30 months for one American electronics company while it was only about 18 months for the Japanese manufacturer of the same product. Clark defines the development cycle as the time taken from the beginning of the product design until the high volume production starts. He observed that the design transfer from development organization to the manufacturing organization was smooth for the Japanese company while for the American company it took several iterations.

The above data, which are indeed limited, together with the Japanese success in capturing consumer electronics, automobiles and other markets prompt us to ask the following question: What makes the Japanese design, development and manufacturing operations so efficient? Admittedly, there are many reasons; but the technological answer to this question is the emphasis placed on design optimization and process optimization for manufacturability, quality and reliability; and the availability of efficient experimentation and simulation methods for performing the optimization.

These methods of design and process optimization were developed in the fifties and early sixties by Genichi Taguchi;[3] Taguchi and Wu.[4] They were developed in an effort to produce high quality products under the post World War

II Japanese circumstances. These circumstances were characterized by low grade raw material, poor manufacturing equipment and inadequate number of well-trained engineers. Now these optimization methods are used routinely to achieve high efficiency in design, development and manufacturing.

In this paper, we will give an overview of the principles involved in design and process optimization for quality, manufacturability and reliability. There are seven sections in this paper. Section 5.2 describes quality engineering and its role in product and manufacturing process design. A general formulation of the design optimization problem is given in Section 5.3; and the optimization strategy is described in Section 5.4. Engineering problems come in a large variety. A broad classification of the engineering problems is given in Section 5.5. In that section, we also describe the selection of the objective function for optimization. The use of orthogonal arrays for evaluating the objective function and for multidimensional optimization is illustrated in Section 5.6 with the help of a temperature controller circuit design. Concluding remarks are presented in Section 5.7.

5.2 QUALITY ENGINEERING

The quality of a product can be quantified in terms of the total loss to the society from the time the product is shipped to the customer Taguchi.[5] The loss may be due to undesirable side effects and due to the deviation of the functional quality from the target performance. Here we will consider only the loss due to functional variation. For example, the amplification level of a public telephone set may differ from cold winter to hot summer; it may differ from one set to another; also it may deteriorate over a period of time. The quadratic loss function can estimate this loss well in most cases. In general a product's performance is influenced by factors which are called noise factors.[1] There are three types of noise factors:

1. *External* - These are various types of noise factors external to the product such as load conditions, temperature, humidity, dust, supply voltage, vibrations due to nearby machinery, and human errors in operating the product.

2. *Manufacturing imperfection* - It includes the variation in the product parameters from unit to unit. This is inevitable in a manufacturing process. An example of the manufacturing imperfection is that the value of a particular resistor in a unit may be specified to be 100 kilo-ohms but in a particular unit it turns out to be 101 kilo-ohms.

---

1. Glossary equivalent is *noise parameter*.

3. *Deterioration* - When the product is sold, all its performance characteristics may be right on target, but as years pass by, the values of individual components may change which lead to product performance deteriorations.

One approach to reduce a product's functional variation is to limit the noise factors or eliminate them altogether. For the telephone set example, it would mean to reduce usable temperature range, to demand tighter manufacturing tolerance or to specify low drift parameters. Undoubtedly, these are costly ways to reduce the public telephone set amplification variation. What is then a less costly way? It is to center the design parameters in such a way as to minimize sensitivity to all noise factors. This involves exploiting the non-linearity of the relationship between the control factors,[2] the noise factors and the response variables. Here, by control factor, we mean the factors or parameters whose levels or values are specified by the designer.

Note that during product design one can make the product robust against all three types of noise factors described above; whereas, during manufacturing process design and actual manufacturing one can reduce variation due to only manufacturing imperfection. Once a product is in the customer's hand, warranty service is the only way to address quality problems. Thus, the major responsibility for the quality of a product lies within the product designers and not with the manufacturing organization.

There are three major steps in designing a product or a manufacturing process: 1) system design, 2) parameter design or design optimization and 3) tolerance design. System design consists of arriving at a workable circuit diagram or manufacturing process layout. The role of parameter design or design optimization is to specify the levels of control factors which minimize sensitivity to all noise factors. During this step, tolerances are assumed to be wide so that manufacturing cost is low. If parameter design fails to produce adequately low functional variation of the product, then during tolerance design tolerances are selectively reduced based upon their cost effectiveness.

Most of the design effort in the United States is concentrated on system design, which involves innovation and the use of engineering sciences, and on tolerance design, which is also known as sensitivity analysis. Design optimization, which is important for producing high quality products at low cost, is ignored. In the remaining part of the paper, we will describe the efficient and easy to use design optimization methods which are commonly used in Japan.

---

2. Glossary equivalent is *control parameter*.

5.3 THE DESIGN OPTIMIZATION PROBLEM

Figure 5.1 shows a block diagram representation of a product. It can also be used to represent a manufacturing process or even a business system. The response is represented by y. The factors which influence the response can be classified into four classes as follows:

1. *Signal factors* (M): These are the factors which are set by the user/operator to attain the target performance or to express the intended output. For example, the steering angle is a signal factor[3] for the steering mechanism of an automobile. The speed control setting on a fan and the bits 0 and 1 transmitted in communication systems are also examples of signal factors. The signal factors are selected by the engineer based on the engineering knowledge. Sometimes even more than one signal factors are used in combination, for example, one signal factor may be used for coarse tuning and one for fine tuning.

2. *Control factors* (z): These are the product parameters specification whose values are the responsibility of the designer.

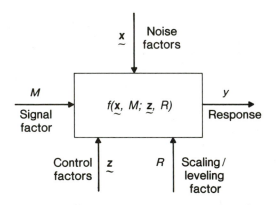

Figure 5.1. Block Diagram of a Product

Each of the control factors can take more than one value which will be referred to as levels. It is the objective of the design activity to determine the best levels of these factors. A number of criteria may be used in defining the best levels; for example, we would want to maximize the stability and

---

3. Glossary equivalent is *signal parameter*.

robustness of the design while keeping the cost minimum. Robustness is the insensitivity to the noise factors.

3. *Scaling/Leveling factors*[4] (R): These are special cases of control factors which can be easily adjusted to achieve desired functional relationship between the signal factor and the response y. For example, the gearing ratio in the steering mechanism can be easily adjusted during the product design phase to achieve desired sensitivity of the turning radius to a change in the steering angle. The threshold voltage in digital communication can be easily adjusted to alter the relative errors of transmitting 0's and 1's.

4. *Noise factors* (x): Noise factors are the uncontrollable factors. They influence the output y and their levels change from one unit of the product to another, from one environment to another and from time to time. Only the statistical characteristics of the noise can be known or specified but not their actual values. Different types of noise factors were described in the section on quality engineering.

Let the dependence and the response y on the signal, control, scaling/leveling and noise factors be denoted by

$$y = f(\mathbf{x}, M; \mathbf{z}, R). \tag{1}$$

Conceptually, the function f consists of two parts: (1) $g(M; \mathbf{z}, R)$ which is the predictable and desirable functional relationship between y and M, and (2) $e(\mathbf{x}, M; \mathbf{z}, R)$ which is the unpredictable and undesirable part.
Thus,

$$y = g(M; \mathbf{z}, R) + e(\mathbf{x}, M; \mathbf{z}, R). \tag{2}$$

In the case where we desire a linear relationship between y and M, g will be a linear function of M. All nonlinear terms will be included in e. Also, the effect of all noise variables is contained in e.

The design objective is to maximize the predictable part and minimize the unpredictable part. This is done by suitably choosing the levels of z and R. As a composite measure of the degree of predictability we will use the ratio of the variances[5] of g and e.

---

4. Glossary equivalent is *adjustment parameter.*

5. By the variances V_g and V_e, we mean the mean square values of $g(M; \mathbf{z}, R)$ and $e(\mathbf{x}, M; \mathbf{z}, R)$, respectively.

That is,

$$\text{Degree of predictability} = \frac{V_g}{V_e}.$$

Note that degree of predictability is a scale invariant quantity. In analogy with communications theory, we take $10 \log_{10}$ of the degree of predictability and call it the signal-to-noise ratio (S/N). The S/N thus defined is a function of both $\mathbf{z}$ and R.

$$\eta'(\mathbf{z},R) = 10 \log_{10} \frac{V_g(\mathbf{z},R)}{V_e(\mathbf{z},R)} \tag{3}$$

As explained earlier, R is a special control factor which is used to achieve desired functional relationship for any given value of $\mathbf{z}$ and to intelligently operate the product in the field so as to maximize the degree of predictability. Further, it is often possible to determine the effect of R on η' without actually having to perform experiments for different values of R. In any case, we first evaluate η ($\mathbf{z}$), where

$$\eta\ (\mathbf{z}) = \max_{R} \eta'(\mathbf{z},R).$$

The design optimization is then performed in the $\mathbf{z}$ domain.

Note that because of the presence of the scaling factor, maximization of η ($\mathbf{z}$) also implies minimization of V_e.[6] Assuming quadratic loss function, minimization of V_e implies minimization of the loss to the customer due to functional variation.

5.4 OPTIMIZATION STRATEGY

The optimization strategy is given by the block diagram of Figure 5.2. The optimization experiments may be conducted in hardware or simulated in software with a mathematical model of the product.

---

6. For a derivation of this result, see: Dehnad, K., and M. Phadke. 1988. "Optimization of Product and Process Design for Quality and Cost." *Quality and Reliability Engineering International* vol. 4. That paper also shows how most of the S/N ratios described in Section 5.5 follow from definition of the S/N ratio given in Equation (3).

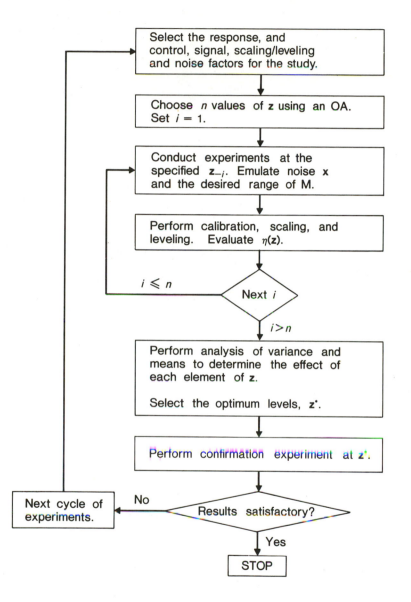

Figure 5.2. Block Diagram for the Optimization Strategy

We begin with selecting the response to be optimized and the four types of factors (control, signal, scaling/leveling and noise) for the study. Using an orthogonal array, we then select a set of n values of the control factors z. For each selected value of z, z_i, we conduct experiments by emulating the noise x and the desired range of M. Orthogonal array experiments provide efficient ways of spanning the noise and signal space. The next step is to perform calibration, scaling and leveling to evaluate the S/N, $\eta (z_i)$. Note that in practice it is not necessary to determine the best value of the scaling/leveling factor, R, for each z_i. It is enough to determine $\eta (z_i)$.

After evaluating $\eta (z_i)$ for $i=1,...,n$ the next step is to perform analysis of means and variance to determine the effect of each element of z on η. This information is then used to select optimum levels of z which are denoted by z^*.

The next step is to perform a confirmation experiment at z^*. If the results are satisfactory, we may terminate the optimization. If not, we reanalyze the data and/or start the next cycle of experiments.

5.5 CLASSIFICATION OF DESIGN PROBLEMS

Design problems can be broadly classified into static and dynamic problems depending upon the absence or presence of signal factors, respectively. There is another major category—life improvement problems. Static and dynamic problems can be further classified depending upon the nature of the signal factor and the response variable. The classification of design problems given below is obviously not complete. However, these classes cover a vast majority of engineering problems encountered in various industries. For some cases, we will derive the S/N in detail while for others we will simply state it.

5.5.1 STATIC PROBLEMS

They are characterized by the absence of signal factors. Depending upon the desired values of the response and whether the response is continuous or discrete, we have the following classes of static problems.

1. *Smaller the better*: Here the response is continuous and positive. Its most desired value is zero. Examples are: the offset voltage of a differential operational amplifier, pollution from a power plant, and leakage current in integrated circuits; the objective function for this case is

$$\eta = -10 \ \log_{10}(Mean \ Squared \ Response).$$

For description of the differential operational amplifier example, see Phadke.[6]

2. *Response on target*: Here the response is continuous and it has a nonextreme target response. Examples are: a) in the window photolithography process of integrated circuit fabrication we would want all contact windows to have the target dimension, 3 microns, say, and b) in making copper cables we would want the wires to have constant resistance per unit length. The S/N for this case is

$$\eta = 10 \, log_{10} \, \frac{\mu^2}{\sigma^2}.$$

Here μ and σ represent the predictable and unpredictable part of the response. For detailed description of the window photolithography application see Phadke, Kackar, Speeney and Grieco.[7]

3. *Larger the better*: This is the case of continuous response where we would like the response to be as large as possible. The strength of a material is an example of this class of problems. The S/N is defined as

$$\eta = -10 \, log_{10} (Mean \ Square \ of \ the \ Reciprocal \ Response).$$

4. *Ordered categorical response*: This is the case where the response is categorized into ordered categories such as very bad, bad, acceptable, good, very good. Accumulation analysis described by Taguchi[8] is the most practical way to analyze such data.

5.5.2 DYNAMIC PROBLEMS

Dynamic problems are characterized by the presence of signal factors. Depending on the nature of the signal factor and the response variable, we have the following four commonly encountered dynamic cases.

1. *Continuous - Continuous* (C-C): Here both the signal factor and the response variable are continuous variables. Design of an analog voltmeter or a temperature controller are examples of the C-C type. In this case, the desired functional relationship is linear. Thus,

$$y(z) = \mu(z,R) + \beta(z,R)M + e(x,M; z,R). \tag{4}$$

Here e also includes the nonlinearity of the relationship between y and M. The parameters μ and β characterize the predictable response. They are determined through calibration experiment and regression analysis. The S/N for the C-C case is defined by

$$\eta = 10 \, log_{10} \, \frac{\beta^2}{\sigma^2}.$$ (5)

Observe that σ^2 is the variance of e which is the unpredictable and undesirable part of y; and β^2 is the variance of the predictable and desirable part except for the range of M.

For the voltmeter design example (in general for all instrument design problems), the S/N can be shown to be directly related to the prediction error. In this problem, M is the applied voltage and y is the needle deflection (Figure 5.3). After establishing Equation (4) through calibration, we would use the following equation to predict the voltage from the observed needle deflection:

$$M = \frac{y-\mu}{\beta} - \frac{e}{\beta}.$$ (6)

The prediction error variance of M is clearly σ^2/β^2. The S/N in Equation (5) is the logarithm of the reciprocal of the prediction error variance.

The problem of design optimization of the C-C case, thus, consists of maximizing η with respect to z. Then, what is the role of R, the scaling/leveling factor? One requirement for the voltmeter design is to get the entire range of voltages represented within the full scale deflection which may be, say, $90°$. This naturally dictates the target value of β. For any z, we can easily adjust R, which would be the gearing ratio, to get the desired β.

Adjustment of this scaling factor R in no way influences η. Thus, it is not necessary to determine the best value of R for each value of z considered

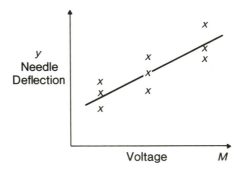

Figure 5.3. Calibration of a Voltmeter

during experimentation. This is a main advantage of the use of S/N as a measure of goodness of the design.

2. *Continuous-Digital* (C-D): A temperature controller where the input (temperature setting) is continuous while the output (the on-off function of the heating unit) is discrete is an example of the C-D case. In such problems, it is preferable to have two separate signal factors, one signal factor (M) for the on function and the other (M') for the off function. For computing the S/N for M, we treat M' as a noise factor while for computing the S/N for M' we treat M as a noise factor.

   By so doing, we can achieve a predictable independent control over both the on and the off functions. In both, M and M', the form of the S/N is same as that for C-C case.

3. *Digital-Continuous* (D-C): The familiar D-A converter is an example of the D-C case. Here again, we use a separate signal factor for the conversion of the 0 and 1 bits. Treatment of the two signal factors is similar to the C-D case.

4. *Digital-Digital* (D-D): Communication systems, computer operations, etc., where both the signal factor and the response are digital are examples of the D-D case. Let us consider a communication system. Table 5.1 shows the detection frequencies for the signal values of 0's and 1's. The probability of wrongly classifying 0 as 1 is p; and the probability of wrongly classifying 1 as 0 is q.

TABLE 5.1. DIGITAL TRANSMISSION TO TABLE

		Output (Response)		
		0	1	Total
Input Signal	0	$n(1-p)$	$n(p)$	n
	1	$n(q)$	$n(1-q)$	n
	Total	$n(1-p+q)$	$n(1+p-q)$	$2n$

The correct detection of 0's and 1's constitute the predictable part of the response. The misclassification is the unpredictable part. By performing analysis of variance, one can compute the sum of squares associated with the predictable (S_g) and unpredictable (S_e) parts as follows:

$$S_g = \frac{n(1-p-q)^2}{2}$$

$$S_e = np(1-p) + nq(1-q)$$

$$S_T = S_g + S_e = \frac{n(1+p-q)(1-p+q)}{2}.$$

Thus the variances associated with the unpredictable and predictable parts are given by

$$V_e = \frac{1}{2n-2} S_e$$

$$V_g = \frac{1}{n} (S_g - V_e).$$

The S/N is then given by

$$\eta = 10 \log_{10} \frac{V_g}{V_e}$$

$$\approx 10 \log_{10} \frac{(1-p-q)^2}{p(1-p) + q(1-q)}. \tag{7}$$

The above approximation assumes that n is large.

Often entropy is used as a measure of the information capacity of a communication system. The advantage of the variance decomposition approach described above lies in that it is easier to generalize it to the case of multiple components of a causal system Taguchi.[3]

Let us now see the role of leveling. It is well known that a communication system is inefficient if the errors of transmitting 0's and 1's are unequal. More efficient transmission is achieved by making $p = q$. This is easily accomplished by the leveling operation which is an operation such as changing the threshold. In Figure 5.4, it can be seen that changing the threshold from R_1 to R_2 balances the two errors and the new system is far more efficient.

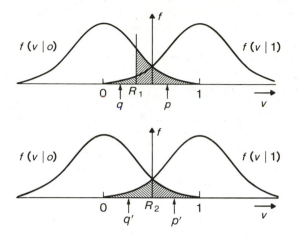

Figure 5.4. Effect of Leveling on Error Probabilities

How to determine $p' = q'$, which are the error rates after leveling, corresponding to the observed error rates p and q? The general formula suggested by Taguchi[3] for approximately determining p' is through the following logit transform equation:

$$2 \; logit \; (p') = logit \; (p) + logit(q)$$

$$2 \times 10 \; log_{10} \left[\frac{1-p'}{p'} \right] = 10 \; log_{10} \left[\frac{1-p}{p} \right] + 10 \; log_{10} \left[\frac{1-q}{q} \right]$$

$$\left[\frac{1}{p'} - 1 \right] = \sqrt{(\frac{1}{p} - 1)(\frac{1}{q} - 1)} . \tag{8}$$

The S/N after leveling is obtained by substituting $p = q = p'$ in Equation (7).

$$\eta = 10 \; log_{10} \frac{(1-2p')^2}{2p'(1-p')} \tag{9}$$

The S/N of Equation (9) is referred as the standardized S/N. It is this S/N that should be maximized with respect to the control factors, z.

In some applications of the D-D case, errors of one kind may be more tolerable than the errors of the other kind. An example of this is the chemical or metallurgical separation. On the input side, 0 would be the metal of interest, say, iron; and 1 would be an impurity. On the output side, 0 would be the molten metal and 1 would be the slag. In this case, the desired ratio of p and q can be determined from cost considerations. Thus, even in this case the logit transform formula can be used to achieve the desired leveling.

5.5.3 LIFE IMPROVEMENT PROBLEMS

The objective of the life improvement problems is to select control factor levels which give maximum life or a desired life distribution. A large number of control factors are studied simultaneously to estimate the effect of each control factor on the survival probability curve. Minute analysis see Taguchi and Wu,[4] which is very similar to accumulation analysis, is used to estimate the survival probability curves and to select the optimum control factor levels. For case studies of life improvements, see Taguchi and Wu[4] and Phadke.[6]

5.6 TEMPERATURE CONTROLLER CIRCUIT DESIGN

5.6.1 THE CIRCUIT DIAGRAM

Figure 5.5 shows a simple circuit diagram for a temperature controller which uses the resistance thermometer, R_T, as the temperature sensor and a variable resistor R_3 to set the desired temperature. The resistance value of R_T at which the relay turns on is given by

$$R_{T-ON} = \frac{R_3 R_2 (E_Z R_4 + E_o R_1)}{R_1 (E_Z R_2 + E_Z R_4 - E_o R_2)}. \tag{10}$$

This problem clearly belongs to the continuous-digital case. We will discuss only the on part of the design.

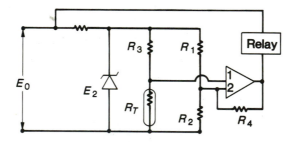

Figure 5.5. Temperature Controller Circuit

5.6.2 SELECTION OF FACTORS

Here the signal factor is $M = R_3$. The five control factors are R_1, R_2, R_4, E_o and E_z. It is convenient to think in terms of the ratios of resistances. So we take R_2/R_1 and R_4/R_1 as control factors rather than R_2 and R_4. Thus $z = (R_1, R_2/R_1, R_4/R_1, E_o, E_z)^T$. For the two voltage parameters, there is a natural restriction that $E_z < E_o$. Table 5.2 lists the starting values of the control factors and their alternate levels.

TABLE 5.2. CONTROL FACTORS AND LEVELS

Label	Factor	Levels		
		1	2	3
A	$R_1(K\Omega)$	2.67	4.0	6.0
B	$\alpha = R_2/R_1$	1.33	2.0	3.0
C	$\lambda = R_4/R_1$	5.33	8.0	16.0
D	E_o	8.0	10.0	12.0
F	E_Z	4.8	6.0	7.2

STARTING
CONDITIONS

The tolerances in the five control factors constitute five noise factors. Environmental factors and deterioration will not be explicitly studied in this analysis. Their effects are assumed to have been adequately represented by the tolerances. The tolerances for the five control factors are given in Table 5.3. Note that the tolerance is specified in terms of the percent of the nominal value for the particular control factor. We assume that the tolerance is equal to three standard deviations.

TABLE 5.3. NOISE FACTORS AND LEVELS

Label	Factor	Tolerance (% of Nominal)	Levels (% of Nominal)		
			1	2	3
A	$R_1(K\Omega)$	5	-2.04	0	2.04
B	$\alpha = R_2/R_1$	5	-2.04	0	2.04
C	$\lambda = R_4/R_1$	5	-2.04	0	2.04
D	E_o	5	-2.04	0	2.04
F	E_z	5	-2.04	0	2.04

5.6.3 S/N EVALUATION

It is inefficient to use Monte Carlo simulation to mimic noise for estimating the S/N. On the contrary, orthogonal arrays can be used to mimic noise in a standardized and efficient way. Here we chose three levels of each noise variable as indicated in Table 5.3. The three levels for each of the noise factors are chosen so that their mean is zero and variance is σ^2, the variance for the noise factor. Thus the three levels are $-\sqrt{3/2}\,\sigma$, 0 and $\sqrt{3/2}\,\sigma$. (Sometimes only two noise levels are considered. They are then taken to be $\pm\sigma$).

We consider three levels for the signal factor: 0.898 kilo-ohm, 1.0 kilo-ohm and 1.102 kilo-ohm. The advantage of taking three levels of the signal factor is that it allows the estimation of the linear and quadratic effects. The linear effect is considered as the signal effect while the quadratic effect is treated as noise.

The L_{18} orthogonal array can be used to study up to seven three-level and one two-level factors. So it is adequate to study the effect of one signal and five noise factors. The L_{18} array and the assignment of columns to noise and signal factors is shown in Table 5.4. Also shown in the last column of Table 5.4 are the $y = R_{T-ON}$ values corresponding to the 18 conditions of the signal and noise factors spelled out by the L_{18} array. For these y values, the control factor settings are: $R_1 = 2.67k\Omega$, $R_2/R_1 = 1.33$, $R_4/R_1 = 5.33$, $E_o = 8.0$ volts and $E_z = 4.8$ volts. That is, each control factor takes the level 1.

Through standard analysis of variance of the 18 y values we can compute the expected values of β^2, which is the square of the linear regression of y on M, and σ_e^2, which is the error variance (see Hicks[9] or Taguchi,[3] for computational formulae). Let M_i be the sum of all y values for which the signal factor is at level i. Also, let T be the sum of all eighteen y values. Then we have,

$$S_\beta = \frac{(-M_1 + M_3)^2}{2\times 6}$$

$$S_T = \sum_{i=1}^{18} y_i^2 - \frac{T^2}{18}$$

$$S_e = S_T - S_\beta$$

$$\eta = 10\, \log_{10} \left\{ \frac{1}{6}\, (S_\beta - \frac{S_e}{18-2}\,)/ \frac{S_e}{18-2} \right\}.$$

For the data in the last column of Table 5.4, we have $\eta = 13.49$. Note that here we treat the distance between two consecutive levels of M (M_2-M_1 and $M_3 - M_1$) as one unit.

TABLE 5.4. L_{18} ORTHOGONAL ARRAY

Expt. No.	1	2	Control Factor Assignment A 3	B 4	5	C 6	D 7	F 8		η
1	1	1	1	1	1	1	1	1	13.49	1.863
2	1	1	2	2	2	2	2	2	13.91	1.892
3	1	1	3	3	3	3	3	3	14.22	1.922
4	1	2	1	1	2	2	3	3	15.29	2.027
5	1	2	2	2	3	3	1	1	15.63	2.154
6	1	2	3	3	1	1	2	2	8.11	2.146
7	1	3	1	2	1	3	2	3	16.46	2.381
8	1	3	2	3	2	1	3	1	-5.52	2.310
9	1	3	3	1	3	2	1	2	16.35	2.281
10	2	1	1	3	3	2	2	1	8.59	1.922
11	2	1	2	1	1	3	3	2	15.77	1.863
12	2	1	3	2	2	1	1	3	15.19	1.892
13	2	2	1	2	3	1	3	2	8.96	2.030
14	2	2	2	3	1	2	1	3	15.52	2.215
15	2	2	3	1	2	3	2	1	15.59	2.087
16	2	3	1	3	2	3	1	2	15.76	2.456
17	2	3	2	1	3	1	2	3	14.82	2.215
18	2	3	3	2	1	2	3	1	9.76	2.309
	1	2	3	4	5	6	7	8		
	M	A	B	C	D	F				y
			Noise Factor Assignment							

5.6.4 OPTIMIZATION

Orthogonal array experimentation is also an efficient way to maximize a non-linear function—in this case to maximize η with respect to the control factors. As mentioned earlier, the L_{18} orthogonal array is adequate to simultaneously study five control factors. The assignment of the control factors to the columns of the L_{18} array is shown in Table 5.4. Using the procedure of Section 5.6.3, we computed S/N for each of the 18 control factor level combinations. The S/N values are listed in the second last column of Table 5.4.

Standard analysis of variance was performed (see Box, Hunter and Hunter[10] or Cochran and Cox[11] on the η values to generate Table 5.5. The mean value for the level 1 of R_1 was computed by averaging the six η values corresponding to 1 in the R_1 column (column number 3). The other mean values were computed similarly. The sum of squares column shows the relative contribution of each factor to the total variance of the 18 η values. The F ratio shows the effect of each factor relative to the error.

From Table 5.5, it is seen that R_1 has little impact on η. The optimum levels for B, C, D and F are respectively B_1, C_3, D_1 and F_3. The S/N corresponding to these optimum control factor settings (A_2, B_1, C_3, D_1 and F_3) is 17.57 db. This represents an improvement of 3.66 db over the S/N for the starting control factor levels (A_2, B_2, C_2, D_2 and F_2).

Further improvement in η can be achieved by investigating lower values of B and higher values of C. Since the two voltages have the natural restriction of $E_z < E_o$, further changes in D and F are not possible. Through two more iterations, the optimum value of η was obtained to be 18.16 db.

TABLE 5.5. ANALYSIS OF VARIANCE for η

Label	Factor	Level Means			Sum of Squares	Deg. of Freedom	Mean Square	F
		1	2	3				
A	R_1	13.09	11.69	13.20	8.56	2	4.28	.5
B	R_2/R_1	15.22	13.32	9.46	103.84	2	51.92	6.6
C	R_4/R_1	9.17	13.24	15.57	125.73	2	62.87	8.0
D	E_o	15.32	12.91	9.75	93.84	2	46.92	5.9
F	E_z	9.59	13.14	15.25	98.18	2	49.09	6.2
	Error				55.30	7	7.90	

The advantages of using orthogonal arrays over many commonly used non-linear programming methods are:

1) no derivatives have to be computed

2) Hessian does not have to be computed

3) algorithm is insensitive to starting conditions

4) large number of variables can be easily handled

5) combinations of continuous and discrete variables can be handled easily.

During design optimization, little attention was paid to the actual R_{T-ON} values as would be needed to control specified temperatures. By selecting of the appropriate middle value for R_3, we can easily customize the design to various temperature ranges. The control factor values we obtained would still be optimum for the various temperature ranges. This is a distinct advantage of using S/N as the objective function. Here adjusting the middle value of R_3 is the scaling/leveling operation.

What if the error variance under the optimum conditions is larger than the desired level? Obviously in this case some of the tolerances would have to be reduced leading to a higher manufacturing cost. To minimize this excess cost, we should first find through analysis of variance the contribution of each noise factor to the error variance and then reduce only those tolerances which give maximum benefit to cost ratio. This is called *tolerance design*.

94

5.7 CONCLUDING REMARKS

Design optimization is essential for producing high quality products at low cost. The role of design optimization is to minimize sensitivity to all noise factors—external noise, manufacturing imperfection and deterioration of parts. Thus, optimization leads to lower manufacturing and operating costs; and it also increases life or reliability. In this paper, we have seen how to formulate the objective function, S/N, for a variety of engineering problems. Also, through the temperature controller circuit example we have shown how orthogonal arrays can be used to evaluate the S/N and maximize it.

Over the last 4 years, many applications of these methods have grown in the United States. Many case studies were reported in the two conferences held this year: Frontiers of Industrial Experimentation Conference organized by AT&T Bell Laboratories and Ford Supplier Symposium on Taguchi Methods sponsored by Ford Motor Company and American Supplier Institute. The application areas include: integrated circuit fabrication, computer aided design of integrated circuits, wave soldering process improvement, tool life optimization and optimization of the operating system of a computer.

ACKNOWLEDGMENTS

The authors would like to thank P. G. Sherry and R. Keny for helpful discussions and comments on this paper.

REFERENCES

1. Juran, J. M. 1981. *Product Quality—A Prescription for the West.* Juran Institute.

2. Clark, K. B. 1984. Series of Management Seminar Given in AT&T Bell Laboratories. April.

3. Taguchi, G. 1976, 1977. *Experimental Designs.* Vol. 1 and 2. Tokyo: Maruzen Publishing Company. (Japanese)

4. Taguchi, G., and Y. Wu. 1980. *Introduction to Off-Line Quality Control.* Tokyo, Japan: Central Japan Quality Control Association.

5. Taguchi, G. 1978. Off-line and On-line Quality Control Systems. *Proceedings of International Conference on Quality Control.* Tokyo, Japan.

6. Phadke, M. S. 1982. "Quality Engineering Using Design of Experiments." *Proceedings of the American Statistical Association, Section on Statistical Education,* 11-20. Cincinnati. (Aug.)

(Quality Control, Robust Design, and
the Taguchi Method; Article 3)

7. Phadke, M. S., R. N. Kackar, D. V. Speeney, and M. J. Grieco. 1983. "Off-line Quality Control in Integrated Circuit Fabrication Using Experimental Design." *The Bell System Technical Journal* **62**: 1273-1309.

(Quality Control, Robust Design, and
the Taguchi Method; Article 6)

8. Taguchi, G. 1974. "A New Statistical Analysis Method for Clinical Data, the Accumulation Analysis, in Contrast with the Chisquare Test." *Saishain-igaku* **29**: 806-813.

9. Hicks, C. R. 1973. *Fundamental Concepts in the Design of Experiments.* New York: Holt, Rinehart and Winston.

10. Box, G. E. P., W. G. Hunter, and J. S. Hunter. 1978. *Statistics for Experimenters—An Introduction to Design, Data Analysis and Model Building.* New York: John Wiley and Sons, Inc.

11. Cochran, W. E., and G. M. Cox. 1957. *Experimental Design.* New York: John Wiley and Sons, Inc.

PART TWO

CASE STUDIES

6

OFF-LINE QUALITY CONTROL IN INTEGRATED CIRCUIT FABRICATION USING EXPERIMENTAL DESIGN

M. S. Phadke, R. N. Kackar, D. V. Speeney, and
M. J. Grieco

6.1 INTRODUCTION AND SUMMARY

This paper describes and illustrates the off-line quality control method, which is a systematic method of optimizing a production process. It also documents our efforts to optimize the process for forming contact windows in 3.5-μm technology complementary metal-oxide semiconductor (CMOS) circuits fabricated in the Murray Hill Integrated Circuit Design Capability Laboratory (MH ICDCL). Here, by optimization we mean minimizing the process variance while keeping the process mean on target.

A typical very large scale integrated circuit (IC) chip has thousands of contact windows (e.g., a BELLMAC[1] -32 microprocessor chip has 250,000 windows on an approximately 1.5-cm^2 area), most of which are not redundant. It is critically important to produce windows of size very near the target dimension. (In this paper windows mean contact windows.) Windows that are not open or are too small result in loss of contact to the devices, while excessively large windows lead to shorted device features. The application of the off-line quality control method has reduced the variance of the window size by a factor of four. Also, it has substantially reduced the processing time required for the window-forming step.

---

1. Trademark of Bell Laboratories.

This study was inspired by Professor Genichi Taguchi's visit to the Quality Theory and Systems Group in the Quality Assurance Center at Bell Laboratories during the months of August, September, and October, 1980. Professor Taguchi, director of the Japanese Academy of Quality and a recipient of the Deming award, has developed the method of off-line quality control during the last three decades. It is used routinely by many leading Japanese industries to produce high-quality products at low cost. An overview of Professor Taguchi's off-line and on-line quality control methods is given in Taguchi[1] and Kackar and Phadke.[2] This paper documents the results of the first application of Professor Taguchi's off-line quality control method in Bell Laboratories.

The distinctive features of the off-line quality control method are experimental design using orthogonal arrays and the analysis of signal-to-noise ratios (s/n). The orthogonal array designs provide an economical way of simultaneously studying the effects of many production factors[2] on the process mean and variance. Orthogonal array designs are fractional factorial designs with the orthogonality property defined in Section 6.4. The s/n is a measure of the process variability. According to Professor Taguchi,[3] by optimizing the process with respect to the s/n, we ensure that the resulting optimum process conditions are robust or stable, meaning that they have the minimum process variation.

The outline of this paper is as follows: Section 6.2 gives a brief description of the window-forming process, which is a critical step in IC fabrication. The window-forming process is generally considered to be one of the most difficult steps in terms of reproducing and obtaining uniform-size windows. Nine key process factors were identified and their potential operating levels were determined. A description of the factors and their levels is given in Section 6.3. The total number of possible factor-level combinations is about six thousand.

The aim of the off-line quality control method is to determine a factor-level combination that gives the least variance for the window size while keeping the mean on target. To determine such a factor-level combination we performed eighteen experiments using the L_{18} orthogonal array. The experimental setup is given in Section 6.4. These eighteen experiments correspond to eighteen factor-level combinations among the possible six thousand combinations. For each experiment, measurements were taken on the line width and the window-size control features. The resulting data were analyzed to determine the optimum factor-level combination. The measurements and the data analysis are presented in Sections 6.5 through 6.9.

The optimum factor levels, inferred from the data analysis, were subsequently used in fabricating the BELLMAC-32 microprocessor, the BELLMAC-4

---

2. Glossary equivalent is *control parameter*.

microcomputer, and some other chips in the Murray Hill ICDCL. The experience of using these conditions is discussed in Section 6.10.

The experiment was designed and preliminary analysis of the experimental data was performed under Professor Taguchi's guidance and collaboration.

6.2 THE WINDOW-FORMING PROCESS

Fabrication of integrated circuits is a complex, lengthy process (Glaser and Subak-Sharp[4]). Window forming is one of the more critical steps in fabricating state of the art CMOS integrated circuits. It comes after field and gate oxides are grown; polysilicon lines have been formed; and the gate, source, and drain areas are defined by the process of doping. Figure 6.1 shows the windows in a cross section of a wafer. A window is a hole of about 3.5 μm diameter etched through an oxide layer of about 2 μm thickness. The purpose of the windows is to facilitate the interconnections between the gates, sources, and drains. For this reason these windows are called contact windows.

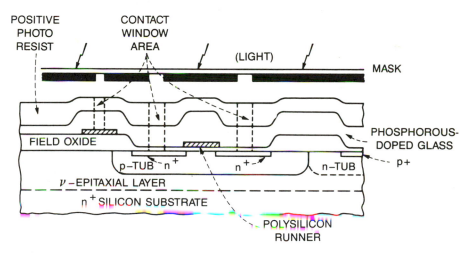

Figure 6.1. Cross Section of Wafer

The process of forming windows through the oxide layers involves photolithography. First the P-glass surface is prepared by depositing undoped oxide on it and prebaking it. The window-forming process is described below.

(i) Apply Photoresist: A wetting agent is sprayed on the wafer to promote adhesion of photoresist to the oxide surface. Then an appropriate photoresist is applied on the wafer and the wafer is rotated at high speed so that the photoresist spreads uniformly.

(ii) Bake: The wafer is baked to dry the photoresist layer. The thickness of the photoresist layer at this stage is about 1.3 to 1.4 μm.

(iii) Expose: The photoresist-coated wafer is exposed to ultraviolet radiation through a mask. The windows to be printed appear as clear areas on the mask. In addition to the windows, which are parts of the desired circuits, the mask has some test patterns. Light passes through these areas and causes the photoresist in the window areas and the test pattern areas to become soluble in an appropriate solvent (developer). The areas of the photoresist where light does not strike remain insoluble.

(iv) Develop: The exposed wafer is dipped in the developer, which dissolves only the exposed areas. In properly printed windows, the exposed photoresist is removed completely and the oxide surface is revealed.

(v) Plasma Etch: The wafers are placed in a high-vacuum chamber wherein a plasma is established. The plasma etches the exposed oxide areas faster than it etches the photoresist. So at the places where the windows are printed, windows are cut through the oxide layers down to the silicon surface.

(vi) Remove Photoresist: The remaining photoresist is now removed with the help of oxygen plasma and wet chemicals.

In the formation of the final contact windows there are additional steps: (vii) removal of cap-oxide, (viii) oxidation of the contact area to prevent diffusion of phosphorus in the subsequent step, (ix) reflow of the P-glass to round the window corners, (x) hydrogen annealing, and (xi) pre-metal wet-etching to remove any remaining oxides from the contact window areas.

At the time we started this study, the target window size at step 6 was considered to be 3.0 μm. The final target window size (after step xi) was 3.5 μm.

6.3 SELECTION OF FACTORS AND FACTOR LEVELS

For the present study only the steps numbered (i) through (v) were chosen for optimization. Discussions with process engineers led to the selection of the following nine factors[3] for controlling the window size. The factors are shown next to the appropriate fabrication steps.

---

3. Glossary equivalent is *control parameter*.

(i) Apply Photoresist: Photoresist viscosity *(B)* and spin speed *(C)*.
(ii) Bake: Bake temperature *(D)* and bake time *(E)*.
(iii) Expose: Mask dimension *(A)*, aperture *(F)*, and exposure time *(G)*.
(iv) Develop: Developing time *(H)*.
(v) Plasma Etch: Etch time *(I)*.

No factor was chosen corresponding to the photoresist removal step because it does not affect the window size.

The standard operating levels of the nine factors are given in Table 6.1. Under these conditions, which prevailed in September 1980, the contact windows varied substantially in size and on many occasions even failed to print and open. The wide variation in window size and the presence of unopened windows is obvious from the figure.

The principle of off-line quality control is to systematically investigate various possible levels for these factors with an aim of obtaining uniform-size windows.

TABLE 6.1. TEST LEVELS

Labels	Factors Name	Levels		
			Standard Levels	
A	Mask Dimension (μm)		2	2.5
B	Viscosity		204	206
C	Spin Speed (rpm)	Low	Normal	High
D	Bake Temperature (°C)	90	105	
E	Bake Time (min)	20	30	40
F	Aperture	1	2	3
G	Exposure Time	20% Over	Normal	20% Under
H	Developing Time (s)	30	45	60
I	Plasma Etch Time (min)	14.5	13.2	15.8

Dependence of spin speed on viscosity

		Spin Speed (rpm)		
		Low	Normal	High
Viscosity	204	2000	3000	4000
	206	3000	4000	5000

Dependence of exposure on aperture

		Exposure (PEP-Setting)		
		20% Over	Normal	20% Under
Aperture	1	96	120	144
	2	72	90	108
	3	40	50	60

In the window-forming experiment a number of alternate levels were considered for each of the nine factors. These levels are also listed in Table 6.1. Six of these factors have three levels each. Three of the factors have only two levels.

The levels of spin speed are tied to the levels of viscosity. For the 204 photoresist viscosity the low, normal, and high spin speeds mean 2000 rpm, 3000 rpm, and 4000 rpm, respectively. For the 206 photoresist viscosity the spin speed levels are 3000 rpm, 4000 rpm, and 5000 rpm. Likewise, the exposure setting depends on the aperture. These relationships are also shown in Table 6.1.

6.4 THE ORTHOGONAL ARRAY EXPERIMENT

The full factorial experiment to explore all possible factor-level combinations would require $3^6 \times 2^3 = 5832$ experiments. Considering the cost of material, the time, and the availability of facilities, the full factorial experiment is prohibitively large. Also from statistical considerations it is unnecessary to perform the full factorial experiment because processes can usually be adequately characterized by a relatively few parameters.

The fractional factorial design used for this study is given in Table 6.2. It is the L_{18} orthogonal array design consisting of 18 experiments taken from Taguchi and Wu.[3] The rows of the array represent runs while the columns represent the factors. Here we treat BD as a joint factor with the levels 1, 2, and 3 representing the combinations B_1D_1, B_2D_1, and B_1D_2, respectively. This is done so that we can study all the nine factors with the L_{18} orthogonal array. Thus, experiment 2 would be run under level 1 of factors A, B, and D, and level 2 of the remaining factors. In terms of the actual settings, these conditions are: 2-μm mask dimension, 204 viscosity, 90°C bake temperature, 3000-rpm spin speed, bake time of 30 minutes, aperture 2, exposure PEP setting 90, 45-second developing time, and 13.2 minutes of plasma etch. The other rows are interpreted similarly.

Here are some of the properties and considerations of this design:

(i)   This is a main-effects-only design; i.e., the response is approximated by a separable function. A function of many independent variables is called separable if it can be written as a sum of functions where each component function is a function of only one independent variable.

(ii)   For estimating the main effects there are two degrees of freedom associated with each three-level factor, one degree of freedom for each two-level factor, and one degree of freedom with the overall mean. We need at least one experiment for every degree of freedom. Thus, the minimum number of experiments needed is $2 \times 6 + 1 \times 3 + 1 = 16$. Our design has 18 experiments. A single-factor-by-single-factor experiment would need only 16 experiments, two fewer than 18. But such an experiment would yield far less precise information compared with the orthogonal array experiment (Taguchi and Wu;[3] Meeker, Hahn, and Feder[5]).

TABLE 6.2. The L_{18} ORTHOGONAL ARRAY

				Column Number & Factor				
Experiment Number	1 A	2 BD	3 C	4 E	5 F	6 G	7 H	8 I
1	1	1	1	1	1	1	1	1
2	1	1	2	2	2	2	2	2
3	1	1	3	3	3	3	3	3
4	1	2	1	1	2	2	3	3
5	1	2	2	2	3	3	1	1
6	1	2	3	3	1	1	2	2
7	1	3	1	2	1	3	2	3
8	1	3	2	3	2	1	3	1
9	1	3	3	1	3	2	1	2
10	2	1	1	3	3	2	2	1
11	2	1	2	1	1	3	3	2
12	2	1	3	2	2	1	1	3
13	2	2	1	2	3	1	3	2
14	2	2	2	3	1	2	1	3
15	2	2	3	1	2	3	2	1
16	2	3	1	3	2	3	1	2
17	2	3	2	1	3	1	2	3
18	2	3	3	2	1	2	3	1

(iii) The columns of the array are pairwise orthogonal. That is, in every pair of columns, all combinations of levels occur and they occur an equal number of time.

(iv) Consequently, the estimates of the main effects of all factors as shown in Table 6.2 and their associated sums of squares are independent under the assumption of normality and equality of error variance. So the significance tests for these factors are independent. Though BD is treated as a joint factor, the main effects and sums of squares of B and D can be estimated separately under the assumption of no interaction. In general, these estimates would be correlated with each other. However, these estimates are not correlated with those for any of the other seven factors.

(v) The estimates of the main effects can be used to predict the response for any combination of the parameter levels. A desirable feature of this design is that the variance of the prediction error is the same for all parameter-level combinations covered by the full factorial design.

(vi) It is known that the main-effect-only models are liable to give mislead-ing conclusions in the presence of interactions. However, in the begin-ning stages of this study the interactions are assumed to be negligible.

If we wished to study all two-factor interactions, with no more than 18 experiments we would have enough degrees of freedom for studying only two three-level factors, or five two-level factors! That would mean in the present study we would have to eliminate half of the process factors without any experimental evidence. Alternately, if we wished to study all the nine process factors and their two-factor interactions, we would need at least 109 experiments! Orthogonal array designs can, of course, be used to study interactions (Taguchi and Wu[3]).

(vii) Optimum conditions obtained from such an experiment have to be verified with an additional experiment. This is done to safeguard us against the potential adverse effects of ignoring the interactions among the manipulatable factors.

In conducting experiments of this kind, it is common for some wafers to get damaged or broken. Also, the wafer-to-wafer variability of window sizes is typically large. So we decided to run each experiment with two wafers.

6.4.1 ANALYSIS OF VARIANCE

Data collected from such experiments are analyzed by a method called analysis of variance (ANOVA) (Hicks[6]). The purpose of ANOVA is to separate the total variability of the data, which is measured by the sum of the squared deviations from the mean value, into contributions by each of the factors and the error. This is analogous to the use of Parseval's theorem to separate the signal strength into contributions by the various harmonics (Taguchi and Wu[3]). To see which of the factors have a significant effect, F-tests are performed. In performing the standard F-test we assume that the errors are normally distributed with equal variance and are independent. The results of the F-test are indicated by the significance level. When we say that a factor is significant at 5-percent level we mean that there is 5 percent or less chance that, if we change the level of the factor, the response will remain the same. If the F-test indicates that a factor is not significant at the 5-percent level it means that, if we change the level of that factor, there is more than a 5-percent chance that the response will remain the same.

The levels of factors which are identified as significant are then set to obtain the best response. The levels of the other factors can be set at any levels within the experimental range. We choose to leave them at the starting levels.

If the assumptions of the F-test are not completely satisfied, the quoted significances are not accurate. However, the standard F-test is relatively insensitive to deviations from the assumptions used in its derivation. Thus, for making engineering decisions about which factor levels to change, the accuracy of the significance level is an adequate guide. In this paper we will use the standard F-test even though some of the assumptions are not strictly satisfied.

6.5 QUALITY MEASURES

The window size is the relevant quality measure for this experiment. The existing equipment does not give reproducible measurements of the sizes of windows in the functional circuits on a chip. This is because of the small size of these windows and their close proximity to one another. Therefore, test patterns—a line-width pattern and a window pattern—are provided in the upper left-hand corner of each chip. The following measurements were made on these test patterns to indicate the quality.

(i) Line width after step (iv), called the pre-etch line width or photo-line width.

(ii) Line width after step (vi), called the post-etch line width.

(iii) Size of the window test pattern after step (vi), called the post-etch window size.

Five chips were selected from each wafer for making the above measurements. These chips correspond to specific locations on a wafer—top, bottom, left, right, and center.

All three quality measures are considered to be good indicators of the size of the functional windows. However, between the geometries of the window-size pattern and the line-width pattern, the geometry of the window-size pattern is closer to the geometry of the functional windows. So, among the three quality measures, the post-etch window size may be expected to be better correlated with the size of the functional windows.

6.6 EXPERIMENTAL DATA

Only thirty-four wafers were available for experimentation. So experiments 15 and 18 were arbitrarily assigned only one wafer each. One of the wafers assigned to experiment 5 broke in handling. So experiments 5, 15, and 18 have only one wafer.

The experimental data are shown in Table 6.3.

The data arising from such experiments can be classified as two types: continuous data and categorical data. Here, the pre-etch and the post-etch line-width data are of the continuous type. The post-etch window size data are mixed categorical-continuous type, because some windows are open while some are not. The two types of data are analyzed somewhat differently, as we explain in the following two sections.

TABLE 6.3. EXPERIMENTAL DATA

Experi-ment No.	Line-Width Control Feature Photoresist—Nanoline Tool (Micrometers)					Comments
	Top	Center	Bottom	Left	Right	
1	2.43	2.52	2.63	2.52	2.5	
1	2.36	2.5	2.62	2.43	2.49	
2	2.76	2.66	2.74	2.6	2.53	
2	2.66	2.73	2.95	2.57	2.64	
3	2.82	2.71	2.78	2.55	2.36	
3	2.76	2.67	2.9	2.62	2.43	
4	2.02	2.06	2.21	1.98	2.13	
4	1.85	1.66	2.07	1.81	1.83	
5	–	–	–	–	–	Wafer Broke
5	1.87	1.78	2.07	1.8	1.83	
6	2.51	2.56	2.55	2.45	2.53	
6	2.68	2.6	2.85	2.55	2.56	
7	1.99	1.99	2.11	1.99	2.0	
7	1.96	2.2	2.04	2.01	2.03	
8	3.15	3.44	3.67	3.09	3.06	
8	3.27	3.29	3.49	3.02	3.19	
9	3.0	2.91	3.07	2.66	2.74	
9	2.73	2.79	3.0	2.69	2.7	
10	2.69	2.5	2.51	2.46	2.4	
10	2.75	2.73	2.75	2.78	3.03	
11	3.2	3.19	3.32	3.2	3.15	
11	3.07	3.14	3.14	3.13	3.12	
12	3.21	3.32	3.33	3.23	3.10	
12	3.48	3.44	3.49	3.25	3.38	
13	2.6	2.56	2.62	2.55	2.56	
13	2.53	2.49	2.79	2.5	2.56	
14	2.18	2.2	2.45	2.22	2.32	
14	2.33	2.2	2.41	2.37	2.38	
15	2.45	2.50	2.51	2.43	2.43	
15	–	–	–	–	–	No wafer
16	2.67	2.53	2.72	2.7	2.6	
16	2.76	2.67	2.73	2.69	2.6	
17	3.31	3.3	3.44	3.12	3.14	
17	3.12	2.97	3.18	3.03	2.95	
18	3.46	3.49	3.5	3.45	3.57	
18	–	–	–	–	–	No wafer

TABLE 6.3. EXPERIMENTAL DATA (Continued)

Experi-ment No.	Line-Width Control Feature Photoresist—Nanoline Tool (Micrometers)					Comments
	Top	Center	Bottom	Left	Right	
1	2.95	2.74	2.85	2.76	2.7	
1	3.03	2.95	2.75	2.82	2.85	
2	3.05	3.18	3.2	3.16	3.06	
2	3.25	3.15	3.09	3.11	3.16	
3	3.69	3.57	3.78	3.55	3.40	
3	3.92	3.62	3.71	3.71	3.53	
4	2.68	2.62	2.9	2.45	2.7	
4	2.29	2.31	2.77	2.46	2.49	
5	–	–	–	–	–	Wafer Broke
5	1.75	1.15	2.07	2.12	1.53	
6	3.42	2.98	3.22	3.13	3.17	
6	3.34	3.21	3.23	3.25	3.28	
7	2.62	2.49	2.53	2.41	2.51	
7	2.76	2.94	2.68	2.62	2.51	
8	4.13	4.38	4.41	4.03	4.03	
8	4.0	4.02	4.18	3.92	3.91	
9	3.94	3.82	3.84	3.57	3.71	
9	3.44	3.30	3.41	3.28	3.20	
10	3.17	2.85	2.84	3.06	2.94	
10	3.70	3.34	3.45	3.41	3.29	
11	4.01	3.91	3.92	3.80	3.90	
11	3.67	3.31	2.86	3.41	3.23	
12	4.04	3.80	4.08	3.81	3.94	
12	4.51	4.37	4.45	4.24	4.48	
13	3.40	3.12	3.11	3.25	3.06	
13	3.22	3.03	2.89	2.92	2.98	
14	3.18	3.03	3.4	3.17	3.32	
14	3.18	2.83	3.17	3.07	3.02	
15	2.86	2.46	2.3	2.6	2.55	
15	–	–	–	–	–	No wafer
16	2.85	2.14	1.22	2.8	3.03	
16	3.4	2.97	2.96	2.87	2.88	
17	4.06	3.87	3.90	3.94	3.87	
17	4.02	3.49	3.51	3.69	3.47	
18	4.49	4.28	4.34	4.39	4.25	
18	–	–	–	–	–	No wafer

TABLE 6.3. EXPERIMENTAL DATA (Continued)

Experi-ment No.	Window-Control Feature Etched—Vickers Tool (Micrometers)					Comments
	Top	Center	Bottom	Left	Right	
1	WNO*	WNO	WNO	WNO	WNO	
1	WNO	WNO	WNO	WNO	WNO	
2	2.32	2.23	2.30	2.56	2.51	
2	2.22	2.33	2.34	2.15	2.35	
3	2.98	3.14	3.02	2.89	3.16	
3	3.15	3.08	2.78	WNO	2.86	
4	WNO	WNO	WNO	WNO	WNO	
4	WNO	WNO	WNO	WNO	WNO	
5	–	–	–	–	–	Wafer Broke
5	WNO	WNO	WNO	WNO	WNO	
6	2.45	2.19	2.14	2.32	2.12	
6	WNO	WNO	WNO	WNO	WNO	
7	WNO	WNO	WNO	WNO	WNO	
7	WNO	WNO	WNO	WNO	WNO	
8	WNO	WNO	WNO	WNO	WNO	
8	2.89	2.97	3.13	3.25	3.19	
9	3.16	2.91	3.12	3.18	3.11	
9	2.43	2.35	2.14	2.40	2.28	
10	2.0	1.75	1.97	1.91	1.72	
10	WNO	2.7	WNO	2.61	2.73	
11	2.76	3.09	3.22	3.05	3.04	
11	3.12	3.21	WNO	2.71	2.27	
12	3.24	3.08	WNO	2.89	2.72	
12	3.5	3.71	3.52	3.53	3.71	
13	2.54	2.63	2.88	2.31	2.71	
13	WNO	WNO	WNO	WNO	WNO	
14	WNO	1.74	2.24	2.07	2.38	
14	WNO	WNO	WNO	WNO	WNO	
15	WNO	WNO	WNO	WNO	WNO	
15	–	–	–	–	–	No wafer
16	WNO	WNO	WNO	WNO	WNO	
16	WNO	WNO	WNO	WNO	WNO	
17	3.09	2.91	3.06	3.09	3.29	
17	3.39	2.5	2.57	2.62	2.35	
18	3.39	3.34	3.45	3.44	3.33	
18	–	–	–	–	–	No wafer

* WNO—Window not open.

6.7 ANALYSIS OF THE LINE-WIDTH DATA

Both the pre-etch and the post-etch widths are continuous variables. For each of these variables the statistics of interest are the mean and the standard deviation. The objective of our data analysis is to determine the factor-level combination such that the standard deviation is minimum while keeping the mean on target. We will call this the optimum factor-level combination. Professor Taguchi's method for obtaining the optimum combination is given next.

6.7.1 SINGLE RESPONSE VARIABLE

Let us first consider the case where there is only one response variable. Instead of working with the mean and the standard deviation, it is preferable to work with the transformed variables—the mean and the signal-to-noise ratio (s/n). The s/n is defined as

$$s/n = \log_{10} \left(\frac{\text{Mean}}{\text{Standard Deviation}} \right)$$

$$= -\log_{10} (\text{coefficient of variation}).$$

In terms of the transformed variables, the optimization problem is to determine the optimum factor levels such that the s/n is maximum while keeping the mean on target. This problem can be solved in two stages:

(i)   Determine which factors have a significant effect on the s/n. This is done through the analysis of variance (ANOVA) of the s/n. These factors are called the *control factors*,[4] implying that they control the process variability. For each control factor we choose the level with the highest s/n as the optimum level. Thus the overall s/n is maximized.

(ii) Select a factor that has the smallest effect on the s/n among all factors that have a significant effect on the mean. Such a factor is called a *signal factor*.[5] Ideally, the signal factor should have no effect on the s/n. Choose the levels of the remaining factors (factors that are neither control factors nor signal factors) to be the nominal levels prior to the optimization experiment. Then set the level of the signal factor so that the mean response is on target.

---

4. Glossary equivalent is *control parameter*.
5. Glossary equivalent is *adjustment parameter*.

In practice, the following two aspects should also be considered in selecting the signal factor: (i) If possible, the relationship between the mean response and the levels of the signal factor should be linear, and (ii). It should be convenient to change the signal factor during production. These aspects are important from the on-line quality control considerations. The signal factor can be used during manufacturing to adjust the mean response (Taguchi;[1] Kackar and Phadke;[2] Taguchi and Wu[3]).

Why do we work in terms of the s/n ratio rather than the standard deviation? Frequently, as the mean decreases, the standard deviation also decreases and vice versa. In such cases, if we work in terms of the standard deviation, the optimization cannot be done in two steps; i.e., we cannot minimize the standard deviation first and then bring the mean on target.

Through many applications, Professor Taguchi has empirically found that the two-stage optimization procedure involving the s/n indeed gives the parameter-level combination where the standard deviation is minimum, while keeping the mean on target. This implies that the engineering systems behave in such a way that the manipulatable production factors can be divided into three categories:

(i) Control factors, which affect process variability as measured by the s/n.

(ii) Signal factors, which do not influence (or have negligible effect on) the s/n but have a significant effect on the mean.

(iii) Factors that do not affect the s/n or the process mean.

The two-stage procedure also has an advantage over a procedure that directly minimizes the mean square error from the target mean value. In practice, the target mean value may change during the process development. The advantage of the two-stage procedure is that for any target mean value (of course, within reasonable bounds) the new optimum factor-level combination is obtained by suitably adjusting the level of only the signal factor. This is so because in step (i) of the algorithm the coefficient of variation is minimized for every mean target value.

6.7.2 MULTIPLE RESPONSE VARIABLES

Now let us consider the case where there are two or more response variables. In such cases, engineering judgment may have to be used to resolve the conflict if different response variables suggest different levels for any one factor. The modified two-stage procedure is as follows:

(i) Separately determine control factors and their optimum levels corresponding to each response variable. If there is a conflict between the optimum levels suggested by the different response variables, use engineering judgment to resolve the conflict.

(ii) Select a factor that has the smallest effect (preferably no effect) on the signal-to-noise ratios for all the response variables but has a significant effect on the mean levels. This is the signal factor. Set the levels of the remaining factors, which affect neither the mean nor the s/n, at the nominal levels prior to the optimization experiment. Then set the level of the signal factor so that the mean responses are on target. Once again engineering judgment may have to be used to resolve any conflicts that arise.

The selection of the control factors, signal factor, and their optimum levels for the present application will be discussed in Section 6.9. The remaining portions of Sections 6.7 and 6.8 contain the data analysis that forms the basis for selecting the optimum factor levels.

6.7.3 PRE-ETCH LINE WIDTH

Mean, standard deviation, and s/n were calculated for each of the eighteen experiments. For those experiments with two wafers, ten data points were used in these calculations. When there was only one wafer, five data points were used. These results are shown in Table 6.4. The presence of unequal sample sizes has been ignored in the subsequent analysis. Let $\bar{x}_i$ and n_i denote the mean and the s/n for the ith experiment.

By computing a single mean $\bar{x}_i$ and a single variance s_i^2 (needed for computing η_i) from the two wafers of each experiment i, we pool together the between wafer and the within wafer variance. That is,

$$E[s_i^2] = \text{(Between wafer variance for experiment } i) \times \frac{5}{9}$$

$$+ \text{(Within wafer variance for experiment } i).$$

Thus, when we maximize η, we minimize the sum of the between-wafer and the within-wafer variances of the line width, which is the response of interest to us. There can be situations when one wants to separately estimate the effects of the factor levels on the between-wafer and within-wafer variances. In those cases, one would compute the s/n and the mean line width for each individual wafer.

In the analysis of the pre-etch and the post-etch line widths, we compute the $\bar{x}_i$ and the s_i^2 for each experiment by pooling the data from both wafers used in that experiment. A relative measure of the between-wafer and within-wafer variance is obtained in Section 6.8, while the post-etch window-size data is being analyzed.

TABLE 6.4. PRE-ETCH LINE-WIDTH DATA

Experiment Number	Mean Line Width, $\bar{x}$ (μm)	Standard Deviation of Line Width, s (μm)	s/n $\eta = \log(\bar{x}/s)$
1	2.500	0.0827	1.4803
2	2.684	0.1196	1.3512
3	2.660	0.1722	1.1889
4	1.962	0.1696	1.0632
5	1.870	0.1168	1.2043
6	2.584	0.1106	1.3686
7	2.032	0.0718	1.4520
8	3.267	0.2101	1.1917
9	2.829	0.1516	1.2709
10	2.660	0.1912	1.1434
11	3.166	0.0674	1.6721
12	3.323	0.1274	1.4165
13	2.576	0.0850	1.4815
14	2.308	0.0964	1.3788
15	2.464	0.0385	1.8065
16	2.677	0.0706	1.5775
17	3.156	0.1569	1.3036
18	3.494	0.0473	1.8692

Analysis of s/n. The estimates of the average s/n for all factor levels are given in Table 6.5. The average for the first level of factor A is the average of the nine experiments (experiments 1 through 9), which were conducted with level 1 of the factor A. Likewise, the average for the second level of factor A is the mean of experiments 10 through 18, which were conducted with level 2 of the factor A. Let us denote these average effects of A_1 and A_2 by mA_1 and mA_2, respectively. Here $mA_1 = 1.2857$ and $mA_2 = 1.5166$. The other entries of Table 6.5 were calculated similarly.

The average signal-to-noise ratios for every level of the eight factors are graphically shown in Figure 6.2. Qualitatively speaking, the mask dimension and the aperture cause a large variation in the s/n. The developing time and the viscosity cause a small change in the s/n. The effect of the other factors is in between.

For three-level factors, Figure 6.2 can also be used to judge the linearity of the effect of the factors. If the difference between levels 1 and 2, and levels 2 and 3 is equal and these levels appear in proper order (1, 2, 3, or 3, 2, 1), then the effect of that factor is linear. If either the differences are unequal or the order is mixed up, then the effect is not linear. For example, the aperture has approximately linear response while the bake time has a nonlinear response.

TABLE 6.5. PRE-ETCH LINE WIDTH FOR AVERAGE S/N

	Factor	Average s/n		
		Level 1	Level 2	Level 3
A	Mask Dimension	1.2857	1.5166	
BD	Viscosity Bake Temperature	(B_1D_1) 1.3754	(B_2D_1) 1.3838	(B_1D_2) 1.4442
B	Viscosity	1.4098	1.3838	
D	Bake Temperature	1.3796	1.4442	
C	Spin Speed	1.3663	1.3503	1.4868
E	Bake Time	1.4328	1.4625	1.3082
F	Aperture	1.5368	1.4011	1.2654
G	Exposure Time	1.3737	1.3461	1.4836
H	Developing Time	1.3881	1.4042	1.4111

Overall average s/n = 1.4011.

We shall perform a formal analysis of variance (ANOVA) to identify statistically significant factors. The analysis of variance of general linear models is widely known in literature, see e.g., Searle[7] and Hicks.[6] Simple ANOVA methods for orthogonal array experiments are described in Taguchi and Wu.[3] The linear model used in analyzing this data is:

$$y_i = \mu + x_i + e_{i,} \qquad (1)$$

where

i = 1, ..., 18 is the experiment number.

μ is the overall mean.

x_i is the fixed effect of the factor-level combination used in experiment i. Here we consider only the main effect for each of the factors. Thus it represents the sum of the effects of the eight factors.

e_i is the random error for experiment i.

y_i is the s/n for experiment i.

To clarify the meaning of the term x_i, let us consider experiment 1, which was run at level 1 of each of the eight factors A through H. Note that the factor I is irrelevant for studying the pre-etch line width. So x_1 is the sum of the main effects associated with the first level of each of the factors A through H.

The sum of squares and the mean squares for the eight factors are tabulated in Table 6.6a. The computations are illustrated in Appendix A.

The expected mean squares are also shown in Table 6.6a. See Hicks[6] and Searle[7] for the computation of expected mean squares, which are used in forming

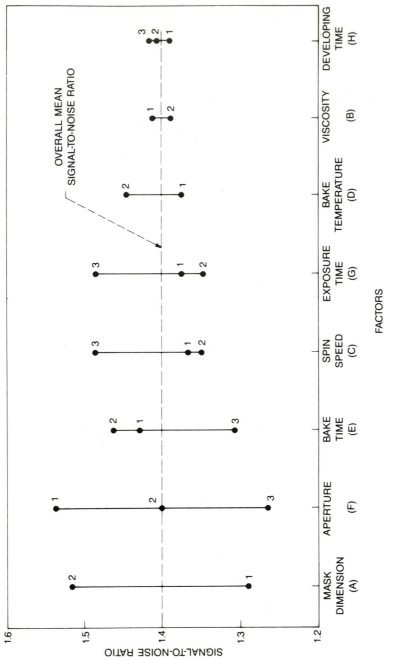

Figure 6.2. **Signal-to-Noise Ratios for Pre-Etch Line Width.** The average s/n for each factor level is indicated by a dot. The number next to the dot indicates the factor level.

appropriate F-tests. The error variance, i.e., variance of e_i, is denoted by σ^2. The variability due to the factors A through H is denoted by ϕ with an appropriate subscript.

In Table 6.6a we see that the mean sum of squares for factors BD, C, G, and H are smaller than the mean error sum of squares. So a new ANOVA table, Table 6.6b, was formed by pooling the sum of squares of these factors with the error sum of squares. The linear model underlying the ANOVA Table 6.6b is the same as Eq. (1), except that now x_i stands for the sum of the main effects of only A, E, and F. The F ratios, computed by dividing the factor mean square by the error mean square, are also shown in Table 6.6b. Factors A and F are significant using F-table values for the 5-percent significance level. So the mask dimension and the aperture are the control factors.

In performing the analysis of variance, we have tacitly assumed that the response for each experiment, here the s/n, has a normal distribution with constant variance. We are presently investigating the distributional properties of the s/n and their impact on the analysis of variance. In this paper we treat the significance levels as approximate.

The engineering significance of a statistically significant factor can be measured in terms of the percent contribution, a measure introduced by Taguchi[3]. The percent contribution is equal to the percent of the total sum of squares explained by that factor after an appropriate estimate of the error sum of squares has been removed from it. The larger the percent contribution, the more can be expected to be achieved by changing the level of that factor. Computation of the percent contribution is illustrated in Appendix B, and the results are shown in Table 6.6b.

From Table 6.6b we see that both the factors A (mask dimension) and F (aperture) contribute in excess of 20 percent each to the total sum of squares. So the factors A and F are not only statistically significant, they have a sizable influence on the s/n. These results are consistent with Figure 6.2. They will be used in Section 6.9 for selecting the control factors.

Analysis of the Means. Now we analyze the mean pre-etch line widths, $\bar{x}_i$ values, to find a signal factor.

The estimates of the mean line widths for all factor levels are given in Table 6.7. These estimates are graphically shown in Figure 6.3. It is apparent that the levels of viscosity, mask dimension, and spin speed cause a relatively large change in the mean line width. Developing time and aperture have a small effect on the line width. The remaining two factors have an intermediate effect.

The linear model used to analyze this data is the same as eq. (1), except that now y_i stands for the mean pre-etch line width rather than the s/n.

The original and the pooled ANOVA tables for the mean pre-etch line width are given in Tables 6.8a and b, respectively. Because the design is not orthogonal with respect to the factors B and D, we need a special method, described in Appendix C, to separate S_{BD} into S_B and S_D.

117

TABLE 6.6. PRE-ETCH LINE WIDTH

(a) ANOVA for s/n

	Source	Degrees of Free-dom	Sum of Squares	Mean Square	Expected Mean Square
A	Mask Dimension	1	0.2399	0.2399	$\sigma^2 + \phi A$
BD	Viscosity Bake Temperature	2	0.0169	0.0085	$\sigma^2 + \phi BD$
C	Spin Speed	2	0.0668	0.0334	$\sigma^2 + \phi C$
E	Bake Time	2	0.0804	0.0402	$\sigma^2 + \phi E$
F	Aperture	2	0.2210	0.1105	$\sigma^2 + \phi F$
G	Exposure Time	2	0.0634	0.0317	$\sigma^2 + \phi G$
H	Developing Time	2	0.0017	0.0009	$\sigma^2 + \phi H$
Error		4	0.1522	0.0381	σ^2
Total		17	0.8423		

(b) Pooled ANOVA for s/n

	Source	Degrees of Free-dom	Sum of Squares	Mean Square	F	Percent Contribution
A	Mask Dimension	1	0.2399	0.2399	9.56*	25.5
E	Bake Time	2	0.0804	0.0402	1.60	3.6
F	Aperture	2	0.2210	0.1105	4.40*	20.3
Error		12	0.3010	0.0251		50.6
Total		17	0.8423			100.00

$F_{1,12}(0.95) = 4.75$. * Factors significant at 95-percent confidence level.
$F_{2,12}(0.95) = 3.89$.

TABLE 6.7. PRE-ETCH LINE WIDTH FOR THE MEAN LINE WIDTH

	Factor	Mean Line Width (μm)		
		Level 1	Level 2	Level 3
A	Mask Dimension	2.39	2.87	
BD	Viscosity Bake Temperature	(B_1D_1) 2.83	(B_2D_1) 2.31	(B_1D_2) 2.74
B	Viscosity	2.79	2.31	
D	Bake Temperature	2.57	2.74	
C	Spin Speed	2.40	2.59	2.89
E	Bake Time	2.68	2.68	2.53
F	Aperture	2.68	2.56	2.64
G	Exposure Time	2.74	2.66	2.49
H	Developing Time	2.60	2.60	2.69

Overall mean line width = 2.63 μm.

It is clear from Table 6.8b that the mask dimension (A), viscosity (B), and spin speed (C) have a statistically significant effect on the mean pre-etch line width. Also, these factors together contribute more than 70 percent to the total sum of squares. These results will be used in Section 6.9 for selecting the signal factor.

TABLE 6.8. PRE-ETCH LINE WIDTH

(a) ANOVA for mean line width

	Source	Degrees of Free-dom	Sum of Squares	Mean Square	Expected Mean Square
A	Mask Dimension	1	1.05	1.050	$\sigma^2 + \phi A$
BD	Viscosity Bake Temperature	2	0.95	0.475	$\sigma^2 + \phi BD$
C	Spin Speed	2	0.73	0.365	$\sigma^2 + \phi C$
E	Bake Time	2	0.10	0.050	$\sigma^2 + \phi E$
F	Aperture	2	0.05	0.025	$\sigma^2 + \phi F$
G	Exposure Time	2	0.19	0.095	$\sigma^2 + \phi G$
H	Developing Time	2	0.04	0.020	$\sigma^2 + \phi H$
Error		4	0.26	0.065	σ^2
Total		17	3.37		

(b) Pooled ANOVA for s/n

	Source	Degrees of Free-dom	Sum of Squares	Mean Square	F	Contribution Percent
A	Mask Dimension	1	1.05	1.050	19.81*	29.6
B	Viscosity	1	0.83	0.834	15.74*	22.6
C	Spin Speed	2	0.73	0.365	6.89*	18.5
G	Exposure Time	2	0.19	0.095	1.79	2.5
Error		11	0.58	0.053		26.8
Total		17	3.37			100.00

$F_{1,11}(0.95) = 4.84.$
$F_{2,11}(0.95) = 3.98.$
* Factors significant at 95-percent confidence level.

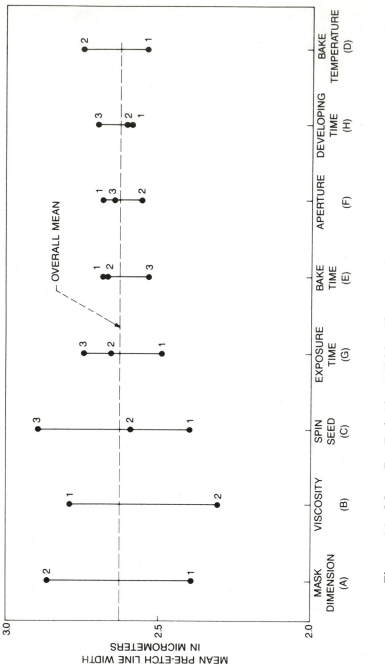

Figure 6.3. Mean Pre-Etch Line Width. The mean line width for each factor level is indicated by a dot. The number next to each dot indicates the factor level.

6.7.4 POST-ETCH LINE WIDTH

The analysis of the post-etch line-width data is similar to the analysis of the pre-etch line-width data. The mean, the standard deviation, and the s/n for each experiment are shown in Table 6.9.

TABLE 6.9. POST-ETCH LINE-WIDTH DATA

Experiment Number	Mean Line Width, $\bar{x}$ (μm)	Standard Deviation of Line Width, s (μm)	s/n $\eta = \log(\bar{x}/s)$
1	2.84	0.11	1.42
2	3.14	0.063	1.70
3	3.65	0.15	1.40
4	2.57	0.20	1.11
5	1.72	0.40	0.63
6	3.12	0.27	1.07
7	2.62	0.19	1.14
8	4.10	0.18	1.37
9	3.55	0.26	1.13
10	3.31	0.35	0.98
11	3.60	0.38	0.98
12	4.17	0.27	1.18
13	3.10	0.16	1.29
14	3.14	0.16	1.29
15	2.55	0.21	1.09
16	2.81	0.37	0.88
18	3.78	0.22	1.23
18	4.34	0.078	1.75

The average s/n and the mean line width for each factor level are shown in Tables 6.10a and b, respectively.

The linear model (1) was again used to analyze the post-etch line-width data. The ANOVA for the signal-to-noise ratios, Table 6.11a, indicates that none of the nine process factors has a significant effect (approximately 5-percent level) on the s/n for the post-etch line width. The pooled ANOVA for the mean post-etch line widths is shown in Table 6.11b. It is obvious from the table that the viscosity, exposure, spin speed, mask dimension, and developing time have significant effects (5-percent level) on the mean line width. The contribution of these factors to the total sum of squares exceeds 90 percent. The mean line width for each factor level is shown graphically in Figure 6.4.

TABLE 6.10. POST-ETCH LINE WIDTH

(a) Average signal-to-noise ratios

		Average s/n		
Factor		Level 1	Level 2	Level 3
A	Mask Dimension	1.22	1.19	
BD	Viscosity Bake Temperature	(B_1D_1) 1.28	(B_2D_1) 1.08	(B_1D_2) 1.25
B	Viscosity	1.27	1.08	
D	Bake Temperature	1.18	1.25	
C	Spin Speed	1.14	1.20	1.27
E	Bake Time	1.16	1.28	1.17
F	Aperture	1.28	1.22	1.11
G	Exposure Time	1.26	1.33	1.02
H	Developing Time	1.09	1.20	1.32
I	Etch Time	1.21	1.18	1.23

Overall Average s/n = 1.205

(b) Mean line width

		Mean Line Width (μm)		
Factor		Level 1	Level 2	Level 3
A	Mask Dimension	3.03	3.42	
BD	Viscosity Bake Temperature	(B_1D_1) 3.45	(B_2D_1) 2.70	(B_1D_2) 3.53
B	Viscosity	3.49	2.70	
D	Bake Temperature	3.08	3.53	
C	Spin Speed	2.88	3.25	3.56
E	Bake Time	3.15	3.18	3.35
F	Aperture	3.28	3.22	3.18
G	Exposure Time	3.52	3.34	2.83
H	Developing Time	3.04	3.09	3.56
I	Etch Time	3.14	3.22	3.32

Overall mean line width = 3.23 μm.

TABLE 6.11. POST-ETCH LINE WIDTH

a) ANOVA for s/n

	Source	Degrees of Freedom	Sum of Squares	Mean Square	F
A	Mask Dimension	1	0.005	0.005	0.02
B	Viscosity	1	0.134	0.134	0.60
D	Bake Temperature	1	0.003	0.003	0.01
C	Spin Speed	2	0.053	0.027	0.12
E	Bake Time	2	0.057	0.028	0.13
F	Aperture	2	0.085	0.043	0.19
G	Exposure Time	2	0.312	0.156	0.70
H	Developing Time	2	0.156	0.078	0.35
I	Etch Time	2	0.008	0.004	0.02
Error		2	0.444	0.222	
Total		17	1.257		

b) Pooled ANOVA for mean line width

	Source	Degrees of Freedom	Sum of Squares	Mean Square	F	Percent Contribution
A	Mask Dimension	1	0.677	0.677	6.92*	8.5
B	Viscosity	1	2.512	2.512	63.51*	32.9
C	Spin Speed	2	1.424	0.712	17.80*	17.9
G	Exposure Time	2	1.558	0.779	19.48*	19.6
H	Developing Time	2	0.997	0.499	12.48*	12.2
Error		9	0.356	0.040		8.9
Total		17	7.524			100.0

$F_{1,9}(0.95) = 5.12.$
$F_{2,9}(0.95) = 4.26.$
* Factors significant at 95-percent confidence level.

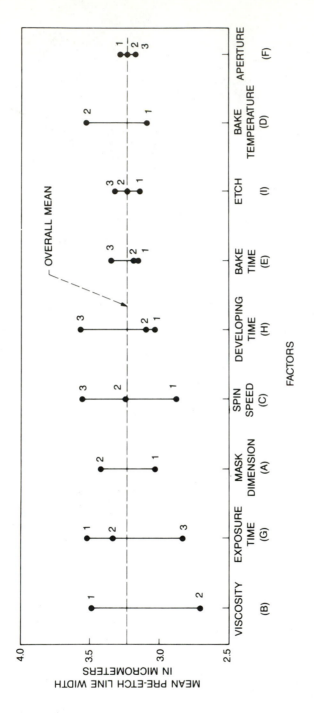

Figure 6.4. Mean Post-Etch Line Width. The mean line width for each factor level is indicated by a dot. The number next to a dot indicates the factor level.

6.8 ANALYSIS OF POST-ETCH WINDOW-SIZE DATA

Some windows are printed and open while the others are not. Thus the window-size data are mixed categorical-continuous in nature. Analysis of such data is done by converting all the data to the categorical type and then using the 'accumulation analysis' method, which is described by Taguchi[1] and Taguchi[8] Factors that are found significant in this analysis are control factors.

The window sizes were divided into the following five categories:

Category	Description (micrometers)	Category	Description (micrometers)
I	Window not open or not printed	III	[2.25, 2.75)
		IV	[2.75, 3.25]
II	(0, 2.25)	V	(3.25, ∞)

Note that these categories are ordered with respect to window size. The target window size at the end of step (vi) was 3 μm. Thus category IV is the most desired category, while category I is the least desired category. Table 6.12 summarizes the data for each of the experiments by categories. To simplify our analysis, we shall presume that a missing wafer has the same readings as the observed wafer for that experiment. This is reflected in Table 6.12, where we show the combined readings for the two wafers of each experiment.

TABLE 6.12. POST-ETCH WINDOW-SIZE DATA—FREQUENCIES BY EXPERIMENT

Experi-ment No.	Frequency Distribution for Wafer 1					Frequency Distribution for Wafer 2					Combined Frequency for the Two Wafers				
	I	II	III	IV	V	I	II	III	IV	V	I	II	III	IV	V
1	5	0	0	0	0	5	0	0	0	0	10	0	0	0	0
2	0	1	0	2	2	0	2	3	0	0	0	3	3	2	2
3	0	0	0	4	0	1	0	0	5	0	1	0	0	9	0
4	5	0	0	0	0	5	0	0	0	0	10	0	0	0	0
5	*	*	*	*	*	5	0	0	0	0	10	0	0	0	0
6	0	3	2	0	0	5	0	0	0	0	5	3	2	0	0
7	5	0	0	0	0	5	0	0	0	0	10	0	0	0	0
8	5	0	0	0	0	0	0	0	5	0	5	0	0	5	0
9	0	0	0	5	0	0	1	4	0	0	0	1	4	5	0
10	0	5	0	0	0	2	0	3	0	0	2	5	3	0	0
11	0	0	0	5	0	1	1	2	1	0	1	1	2	6	0
12	1	0	1	3	0	0	0	0	0	4	1	0	1	3	5
13	0	0	3	2	0	5	0	0	0	0	5	0	3	2	0
14	1	3	1	0	0	5	0	0	0	0	6	3	1	0	0
15	5	0	0	0	0	*	*	*	*	*	10	0	0	0	0
16	5	0	0	0	0	5	0	0	0	0	10	0	0	0	0
17	0	0	0	3	2	0	0	4	0	1	0	0	4	3	3
18	0	0	0	0	5	*	*	*	*	*	0	0	0	0	10

* Implies data missing.

Table 6.13 gives the frequency distribution corresponding to each level of each factor. To obtain the frequency distribution for a specific level of a specific factor, we summed the frequencies of all the experiments that were conducted with that particular level of that particular factor. For example, the frequency distribution for the first level of factor C (low spin speed) was obtained by summing the frequency distributions of experiments with serial numbers 1, 4, 7, 10, 13, and 6. These six experiments were conducted with level 1 of factor C.

TABLE 6.13. POST-ETCH WINDOW-SIZE DATA—FREQUENCIES BY FACTOR LEVEL

Factor Levels	Frequencies					Cumulative Frequencies				
	I	II	III	IV	V	(I)	(II)	(III)	(IV)	(V)
Mask Dimension										
A_1	51	7	9	21	2	51	58	67	88	90
A_2	35	9	14	14	18	35	44	58	72	90
Viscosity, Bake Temperature										
B_1D_1	15	9	9	20	7	15	24	33	53	60
B_2D_1	46	6	6	2	0	46	52	58	60	60
B_1D_2	25	1	8	13	13	25	26	34	47	60
Spin Speed										
C_1	47	5	6	2	0	47	52	58	60	60
C_2	22	7	10	16	5	22	29	39	55	60
C_3	17	4	7	17	15	17	21	28	45	60
Bake Time										
E_1	31	2	10	14	3	31	33	43	57	60
E_2	26	3	7	7	17	26	29	36	43	60
E_3	29	11	6	14	0	29	40	46	60	60
Aperture										
F_1	32	7	5	6	10	32	39	44	50	60
F_2	36	3	4	10	7	36	39	43	53	60
F_3	18	6	14	19	3	18	24	38	57	60
Exposure Time										
G_1	26	3	10	13	8	26	29	39	52	60
G_2	18	12	11	7	12	18	30	41	48	60
G_3	42	1	2	15	0	42	43	45	60	60
Developing Time										
H_1	37	4	6	8	5	37	41	47	55	60
H_2	27	11	12	5	5	27	38	50	55	60
H_3	22	1	5	22	10	22	23	28	50	60
Etch Time										
I_1	37	5	3	5	10	37	42	45	50	60
I_2	21	8	14	15	2	21	29	43	58	60
I_3	28	3	6	15	8	28	31	37	52	60
Totals	86	16	23	35	20	86	102	125	160	180

The frequency distributions of Table 6.13 are graphically displayed by star plots in Figure 6.5. From this figure and the table it is apparent that a change in the level of viscosity, spin speed, or mask dimension causes a noticeable change in the frequency distribution. A change in the level of etch time, bake time, or bake temperature seems to have only a small effect on the frequency distribution. The effects of the other factors are intermediate.

We now determine which factors have a significant effect on the frequency distribution of the window sizes. The standard chi-square test for multinomial distributions is not appropriate here because the categories are ordered. The accumulation analysis method has an intuitive appeal and has been empirically found by Professor Taguchi to be effective in analyzing ordered categorical data. The method consists of the following three steps:

(i) Compute the cumulative frequencies. Table 6.13 shows the cumulative frequencies for all factor levels. The cumulative categories are denoted with parentheses. Thus (III) means sum of categories I, II, and III. Note that the cumulative category (V) is the same as the total number of window readings for the particular factor level.

(ii) Perform "binary data" ANOVA[7] on each cumulative category except the last category, viz. (V). Note that a certain approximation is involved in the significance level suggested by this ANOVA because the observations are not normally distributed.

(iii) Assign weights to each cumulative category. These weights are inversely proportional to the Bernoulli trial variance. Let cum_c be the total number of windows in the cumulative category, c, as given in the bottom row of Table 6.13.

Then the weight for that category is:

$$W_c = \frac{1}{\frac{cum_c}{180} \times \left[1 - \frac{cum_c}{180}\right]} = \frac{180^2}{cum_2(180 - cum_c)},$$

These weights are shown in Appendix D for each category.

Then for each factor and for each error term the accumulated sum of squares is taken to be equal to the weighted sum of the sum of squares for all cumulative categories.

The intuitive appeal for accumulation analysis is that by taking cumulative frequencies we preserve the order of the categories. By giving weights inversely proportional to the sampling errors in each cumulative category, we make the procedure more sensitive to a change in the variance. The difficulty is that the frequencies of the cumulative categories are correlated. So the true level of

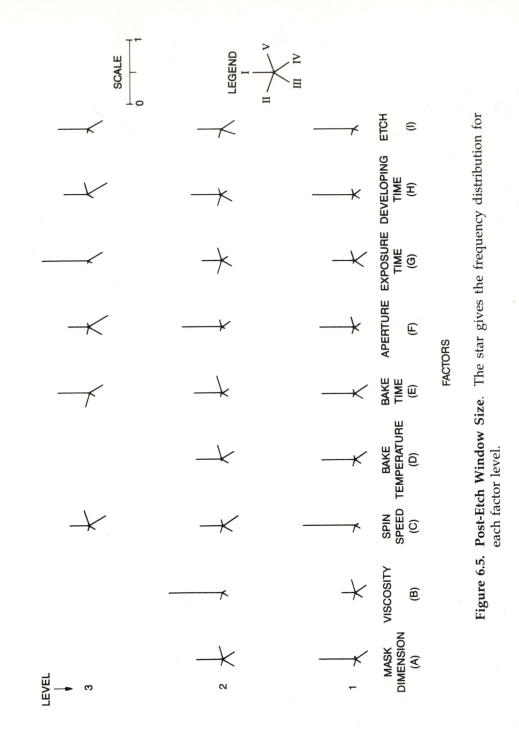

Figure 6.5. Post-Etch Window Size. The star gives the frequency distribution for each factor level.

significance of the F-test may be somewhat different from that indicated by the F table. More work is needed to understand the statistical properties of the accumulation analysis.

Table 6.14 gives the final ANOVA with accumulated sum of squares. The computations are illustrated in Appendix D. For each cumulative category, the following nested, mixed linear model was used in performing the ANOVA:

$$y_{ijk} = \mu + x_i + e_{1ij} + e_{2ijk},$$ (2)

where

$i = 1, ..., 18$ stands for the experiment.

$j = 1, 2$ stands for wafer within the experiment.

$k = 1, ..., 5$ stands for replicate or position within wafer within the experiment.

μ is the overall mean.

x_i is the fixed effect of the factor-level combination used in experiment i. Here we consider only the main effect for each of the factors. See the discussion of model (1) in Section 6.7.3 for more details of the interpretation of x_i.

e_{1ij} is the random effect for wafer j within experiment i.

e_{2ijk} is the random error for replicate k within wafer j within experiment i.

y_{ijk} is the observation for replicate k in wafer j in experiment i. y_{ijk} takes a value 1 if the window size belongs to the particular category. Otherwise, the value is zero.

The expected mean squares for this ANOVA model are also shown in Table 6.14. The variances of e_1 and e_2 are denoted by σ_1^2 and σ_2^2, respectively. The effects of the factors A through I are denoted by ϕ with an appropriate subscript. The effect of lack of fit is denoted by ϕ_l. We assume that the random variables e_{1ij} and e_{2ijk} are independent for all values of i, j and k. The degrees of freedom shown in Table 6.14 have been adjusted for the fact that three experiments have only one wafer each.

For testing the significance of the effect of error between wafers within experiments, the relevant denominator sum of squares is the estimate of σ_2^2. The corresponding F value is 11.69, which is significant far beyond the nominal 5-percent level. To test for the lack of fit of the main-effects-only model, the appropriate denominator is the estimate $\sigma_2^2 + 5\sigma_1^2$. The corresponding F ratio is 0.87. This indicates that the main-effects-only model adequately describes the observed data relative to the random errors between wafers. For testing the significance of the process factors, the denominator mean square is again the

129

estimate of $\sigma_2^2 + 5\sigma_1^2$. We see that the mask dimension, viscosity, spin speed, exposure time, and developing time have a significant effect (approximately 5-percent level) on the window size. The effects of the other factors are not significant.

TABLE 6.14. POST-ETCH WINDOW SIZE

(a) ANOVA for accumulation analysis

Source	Degrees of Free- dom	Sum of Squares	Mean Square	F	(Expected Mean Square) ÷ $\overline{W}$
A Mask Dimension	4	26.64	6.66	2.67*	$\sigma_2^2 + 5\sigma_1^2 + \phi A$
BD Viscosity-Bake Temperature	8	112.31	14.04	5.64*	$\sigma_2^2 + 5\sigma_1^2 + \phi BD$
C Spin Speed	8	125.52	15.69	6.30*	$\sigma_2^2 + 5\sigma_1^2 + \phi C$
E Bake Time	8	36.96	4.62	1.86	$\sigma_2^2 + 5\sigma_1^2 + \phi E$
F Aperture	8	27.88	3.49	1.40	$\sigma_2^2 + 5\sigma_1^2 + \phi F$
G Exposure Time	8	42.28	5.29	2.12*	$\sigma_2^2 + 5\sigma_1^2 + \phi G$
H Developing Time	8	45.57	5.70	2.29*	$\sigma_2^2 + 5\sigma_1^2 + \phi H$
I Etch Time	8	23.80	2.98	1.20	$\sigma_2^2 + 5\sigma_1^2 + \phi I$
Lack of Fit	8	17.25	2.16	0.87	$\sigma_2^2 + 5\sigma_1^2 + \phi \iota$
Error Between Wafers Within Experiment	60	149.33	2.49	11.69*	$\sigma_2^2 + 5\sigma_1^2$
Error Between Repli- cates Within Wafers Within Experiment	528	112.45	0.21		σ_2^2
Total	656	720.00			

$\overline{W} = (W_{(I)} + W_{(II)} + W_{(III)} + W_{(IV)})/4$.
$F_{4,60}(0.95) + 2.53, F_{8,60}(0.95) = 2.10, F_{60,528}(0.95) = 1.32$.

(b) Separation of S_{BD}

Source	Degrees of Free- dom	Sum of Squares	Mean Square	F
B Viscosity	4	87.38	21.85	8.78*
D Bake Temperature	4	6.55	1.64	0.66

* Factors significant at 95-percent confidence level.

6.9 SELECTION OF OPTIMUM FACTOR LEVELS

The following table summarizes the significant results of the analyses performed in Sections 6.7 and 6.8. In each category, the factors are arranged in descending order according to the F value.

Significant effect on s/n:
  Pre-etch line width: A, F
  Post-etch line width: None
Significant effect on mean:
  Pre-etch line width: A, B, C
  Post-etch line width: B, G, C, A, H
Significant factors identified by accumulation analysis:
  Post-etch window size: B, C, A, H, G

Factors that have a significant effect on the s/n and the factors identified to be significant by the accumulation analysis are all control factors. Setting their levels equal to optimum levels minimizes the process variability. Here the control factors are A, F, B, C, H, and G.

To keep the process mean on target we use a signal factor. Ideally, the signal factor should have a significant effect on the mean, but should have no effect on the s/n. Then changing the level of the signal factor would affect only the mean. In practice, a small effect on the s/n may have to be tolerated.

Among the factors (A, B, C, G, and H) that have a significant effect on the mean, factors A, B, and C are relatively strong control factors as measured by the F statistics for the accumulation analysis and the ANOVA for pre-etch line-width s/n. Also, these factors are relatively difficult to change during production. So A, B, and C are not suitable as signal factors. Between the remaining two factors, G and H, G has greater effect on the mean and also shows as a less significant factor in accumulation analysis. So exposure time was assigned to be the signal factor.

The optimum levels for the control factors were selected as follows. The mask dimension (A) and the aperture (F) have a significant effect on the s/n for pre-etch line width. From Table 6.5 we see that the 2.5-µm mask (level 2) has a higher s/n than the 2.0-µm mask. Hence 2.5 µm was chosen to be the optimum mask dimension. Also, aperture 1 (level 1) has the highest s/n among the three apertures studied. However, because of the past experience, aperture 2 was chosen to be the preferred level.

The accumulation analysis of the post-etch window-size data indicated that the viscosity, spin speed, mask dimension, developing time, and exposure have statistically significant effects on the frequency distribution. The optimum levels of these factors can be determined from Table 6.13 and Figure 6.5 to be those that have the smallest fraction of windows not open (category I) and the largest fraction of windows in the range 3.0 ± 0.25 µm (category IV). Because it is more critical to have all the windows open, when there was a conflict we took the requirement on category I to be the dominant requirement. The optimum levels are: 2.5-µm mask dimension, viscosity 204, 4000-rpm spin speed, 60-second developing time, and normal exposure.

Table 6.15 shows side by side the optimum factor levels and the standard levels as of September 1980. Note that our experiment has indicated that the mask dimension be changed from 2.0 µm to 2.5 µm, spin speed from 3000 rpm to 4000 rpm, and developing time from 45 seconds to 60 seconds. The exposure time is to be adjusted to get the correct mean value of the line width and the window size. The levels of the other factors, which remain unchanged, have been confirmed to be optimum to start with.

In deriving the optimum conditions we have conducted a highly fractionated factorial experiment and have considered only the main effects of the factors. The interactions between the factors have been ignored. If the interactions are strong compared to the main effects, then there is a possibility that the optimum conditions thus derived would not improve the process. So experiments have to be conducted to verify the optimum conditions. The verification was done in conjunction with the implementation, which is described next.

TABLE 6.15. OPTIMUM FACTOR LEVELS

Label	Factors Name	Standard Levels	Optimum Levels
A	Mask Dimension (µm)	2.0	2.5
B	Viscosity	204	204
C	Spin Speed (rpm)	3000	4000
D	Bake Temperature (°C)	105	105
E	Bake Time (min)	30	30
F	Aperture	2	2
G	Exposure (PEP setting)	Normal	Normal
H	Developing Time (s)	45	60
I	Plasma Etch Time (min)	13.2	13.2

6.10 IMPLEMENTATION AND THE BENEFITS OF THE OPTIMUM LEVELS

We started to use the optimum process conditions given in Table 6.15 in the Integrated Circuits Design Capability Laboratory in January 1981. In the beginning the exposure was set at 90, which is the normal setting given in Table 6.1. We observed that the final window at the end of step *(xi)* was much larger than the target size of 3.5 µm. Through successive experiments, we reduced the exposure time until the mean final window size came to about 3.5 µm. The corresponding exposure setting is 140. Since then the process has been run at these conditions. The benefits of running the process at these conditions are:

(i) The pre-etch line width is routinely used as a process quality indicator. Before September 1980 the standard deviation of this indicator was 0.29 µm on a base line chip (DSO chip). With the optimum process parameters, the standard deviation has come down to 0.14 µm. This is a two-fold reduction in standard deviation, or a four-fold reduction in variance. This was evidenced by a typical photograph of the PLA area of a BELLMAC-32 microprocessor chip fabricated by the new process. The windows in that photograph were much more uniform in size than those in previous photographs. Also, all windows were printed and opened.

(ii) After the final step of window forming, i.e., after step *(xi)*, the windows are visually examined on a routine basis. Analysis of the quality control data on the DSO chip, which has an area of approximately 0.19 *cm²*, showed that prior to September 1980 about 0.12 window per chip was either not open or not printed (i.e., approximately one incidence of window not open or not printed was found in eight chips). With the new process only 0.04 window per chip is not open or printed (i.e., approximately one incidence of window not open or printed is found in twenty-five chips). This is a three-fold reduction in defect density due to unopened windows.

(iii) Observing these improvements over several weeks, the process engineers gained a confidence in the stability and robustness of the new process parameters. So they eliminated a number of in-process checks. As a result the overall time spent by wafers in window photolithography has been reduced by a factor of two.

The optimum parameter levels were first used in the Integrated Circuit Device Capability Laboratory with only a few codes of ICs. Subsequently, these parameter levels were used with all codes of 3.5-µm technology chips, including BELLMAC-4 microcomputer and BELLMAC-32 microprocessor chips. The mask dimension change from 2.0 to 2.5 µm is now a standard for 3.5-µm CMOS technology.

6.11 DISCUSSION AND FUTURE WORK

The off-line quality control method is an efficient method of improving the quality and the yield of a production process. The method has a great deal of similarity with the response surface method (Myers[9]) and the evolutionary operations method (Box and Draper[10]), which are commonly known in statistical literature in this country. Both the response surface and the evolutionary operations methods are used to maximize the yield of a production process and they both make use of the experimental design techniques. The main difference is that in the off-line quality control method the process variability that has a great impact on the product quality is the objective function. In the response surface and evolutionary operations methods, the process variability is generally not considered. Thus, intuitively, the optimum levels derived by using the off-line quality control method can be expected to be more robust, stable, and dependable.

In the response surface method one typically uses a relatively large fraction of the factorial experiment. However, in off-line quality control usually a very small fraction is chosen. Another difference is that in the response surface method the objective function is considered to be a continuous function approximated by a low-order polynominal. In off-line quality control, we can simultaneously study both the continuous and discrete factors.

Our application of the off-line quality control method to the window-cutting process in the Murray Hill 3.5-μm CMOS technology, as seen from the earlier sections, has resulted in improved control of window size, lower incidence of unopened windows, and reduced time for window photolithography. Presently, we have undertaken to optimize two more steps in IC fabrication. Those steps are polysilicon patterning and aluminum patterning. Both these processes, like the window-cutting process, involve photolithography and are among the more critical processes of IC fabrication. We think that the method has a great potential and would like to see applications in various parts of Bell Laboratories and Western Electric Company.

ACKNOWLEDGMENTS

The authors acknowledge with gratitude the support and the encouragement given by R. L. Chaddha, R. Edwards, J. V. Dalton, A. B. Hoadley, and G. R. Weber. R. G. Brandes, K. J. Orlowski, and members of the IC Process Operations Group provided valuable help in conducting the experiments. We also thank Ramon Leon for helpful discussions on the graphical representation of the data and I. J. Terpenning for helpful comments on an earlier draft of this paper. The original manuscript was carefully prepared by Jill Cooper on the UNIX[6] operating system.

---

6. Registered trademark of AT&T.

APPENDIX A

COMPUTATION OF THE SUM OF SQUARES—ANALYSIS OF THE SIGNAL-TO-NOISE RATIO FOR PRE-ETCH

The computations of the sum of squares tabulated in Table 6.6a are illustrated below.

$$S_m = \text{Correction Factor}$$

$$= \frac{\left[\sum\limits_{i=1}^{18} \eta_i\right]^2}{18} = \frac{(25.2202)^2}{18} = 35.3366$$

$$S_A = \text{Sum of squares for factor A}$$

$$= \frac{(9m_{A_1})^2 + (9m_{A_2})^2}{9} - Sm$$

$$= \frac{(11.5711)^2 + (13.6491)^2}{9} - 35.3366$$

$$= 0.2399 \qquad (d.f.=1)$$

$$S_c = \frac{(6m_{C_1})^2 + (6m_{C_2})^2 + (6m_{C_3})^2}{6} - Sm$$

$$= \frac{(8.1979)^2 + (8.1017)^2 + (8.9206)^2}{6} - S_m$$

$$= 0.0668 \qquad (d.f.=2).$$

Sums of squares for the factors E, F, G, and H were calculated similarly. The combined sum of squares due to B and D is given by

$$S_{BD} = \text{Sum of squares for the column BD}$$

$$= \frac{(6m_{B_1D_1})^2 + (6m_{B_1D_2})^2 + (6m_{B_3D_1})^2}{6} - S_m$$

$$= \frac{(8.2524)^2 + (8.6649)^2 + (8.3029)^2}{6} - 35.3366$$

$$= 0.0169 \qquad (d.f.=2).$$

The total sum of squares is

$$S_T = \sum_{i=1}^{18} \eta_i^2 - S_m = 0.8423. \qquad (d.f.=17).$$

The error sum of squares is calculated by subtraction.

$$S_e = S_T - (S_A + S_{BD} + S_C + S_E + S_F + S_G + S_H)$$

$$= 0.1522 \qquad (d.f.=4).$$

Here we do not compute the sum of squares due to factor I (etch time), because it has no influence on the pre-etch line width.

APPENDIX B

COMPUTATION OF THE PERCENT CONTRIBUTION—ANALYSIS OF THE SIGNAL-TO-NOISE RATIO FOR PRE-ETCH LINE WIDTH

The computation of the percent contribution is explained below. The contribution of factor A to the total sum of squares

$$= S_A - (d.f. \text{ of A})(\text{error mean square}).$$

Hence, the percent contribution for factor A

$$= \frac{S_A - (\text{d.f. of A})(\text{error mean square})}{\text{total sum of squares}} \times 100$$

$$= \frac{0.2399 - 0.0251}{0.8423} \times 100 = 25.5\%.$$

The percent contributions of E and F are determined similarly. Now consider, the contribution of error to the total sum of squares:

$$= S_e + (\text{total d.f. for factors })(\text{error mean square}).$$

Hence, the percent contribution for error

$$= \frac{S_e + (\text{total d.f. for factors})(\text{error mean square})}{\text{total sums of squares}} \times 100$$

$$= \frac{0.3010 + 5 \times 0.0251}{0.8423} \times 100 = 50.6\%.$$

APPENDIX C

SEPARATION OF S_{BD} INTO S_B AND S_D—ANALYSIS OF THE MEAN PRE-ETCH LINE WIDTH

The sum of squares, S_{BD}, can be decomposed in the following two ways (Snedecor and Cochran[11]) to obtain the contributions of the factors B and D:

$$S_{BD} = S'_{B(D)} + S'_D$$

and

$$S_{BD} = S'_{D(B)} + S'_B.$$

Here $S'_{D(B)}$ is the sum of squares due to B, assuming D has no effect; S'_D is the sum of squares due to D after eliminating the effect of B. The terms $S'_{D(B)}$ and S'_B are interpreted similarly. We have

$$S'_{B(D)} = \frac{(6m_{B_1D_1} + 6m_{B_1D_2} - 12m_{B_2D_1})^2}{(1^2 + 1^2 + 2^2) \times 6} = 0.903 \qquad (d.f. = 1)$$

$$S'_D = S_{BD} - S'_{B(D)} = 0.047 \qquad (d.f. = 1).$$

Similarly,

$$S'_{D(B)} = \frac{(6m_{B_1D_1} + 6m_{B_2D_1} - 12m_{B_1D_2})^2}{(1^1 + 1^2 + 2^2) \times 6} = 0.116 \qquad (d.f. = 1)$$

$$S'_B = S_{BD} - S'_{D(B)} = 0.834 \qquad (d.f. = 1).$$

For testing the significance of the factors B and D we use S'_B and S'_D, respectively. Note that S'_B and S'_D do not add up to S_{BD}, which is to be expected because the design is not orthogonal with respect to the factors B and D.

APPENDIX D

COMPUTATION OF THE SUM OF SQUARES FOR ACCUMULATION ANALYSIS—ANALYSIS OF POST-ETCH WINDOW SIZE

The weights for the cumulative categories (I), (II), (III), and (IV) are given below. The frequencies of the bottom line of Table 6.13 are used in computing these weights. Therefore:

$$W_{(I)} = \frac{1}{\dfrac{86}{180} \times \dfrac{180-86}{180}} = 4.008$$

$$W_{(II)} = \frac{180^2}{102 \times (180-102)} = 4.072$$

$$W_{(III)} = \frac{180^2}{125 \times (180-125)} = 4.713$$

$$W_{(IV)} = \frac{180^2}{160 \times (180-160)} = 10.125.$$

Computation of the sum of squares tabulated in Table 6.14 are illustrated below:

$$S_A = W_{(I)} \times \left[\frac{51^2 + 35^2}{90} - \frac{86^2}{180} \right] + W_{(II)} \times \left[\frac{58^2 + 44^2}{90} - \frac{102^2}{180} \right]$$

$$+ W_{(III)} \times \left[\frac{67^2 + 58^2}{90} - \frac{125^2}{180} \right] + W_{(IV)} \times \left[\frac{88^2 + 72^2}{90} - \frac{160^2}{180} \right] = 26.64,$$

139

and,

$$S_C = W_{(I)} \times \left[\frac{47^2 + 22^2 + 17^2}{60} - \frac{86^2}{180} \right]$$

$$+ W_{(II)} \times \left[\frac{52^2 + 29^2 + 21^2}{60} - \frac{102^2}{180} \right]$$

$$+ W_{(III)} \times \left[\frac{58^2 + 39^2 + 28^2}{60} - \frac{125^2}{180} \right]$$

$$+ W_{(IV)} \times \left[\frac{60^2 + 55^2 + 45^2}{90} - \frac{160^2}{180} \right]$$

$$= 125.52.$$

REFERENCES

1. Taguchi, G. 1978. Off-line and On-line Quality Control Systems. *International Conference on Quality Control*, B4-1-5. Tokyo, Japan.

2. Kackar, R. N., and M. S. Phadke. n.d. *An Introduction to Off-line and On-line Quality Control Methods*. Unpublished work.

3. Taguchi, G., and Y. Wu. 1980. *Introduction to Off-Line Quality Control*. Nagoya, Japan: Central Japan Quality Control Association.

4. Glaser, A. B., and G. A. Subak-Sharp. 1977. *Integrated Circuits Engineering*. Reading, Mass.: Addison-Wesley.

5. Meeker, W. Q., G. H. Hahn, and P. I. Feder. 1975. A Computer Program for Evaluating and Comparing Experimental Designs and Some Applications. *American Statistician* **29** (Feb. no. 1):60-64.

6. Hicks, C. R. 1973. *Fundamental Concepts in the Design of Experiments*. New York: Holt, Rinehart and Winston.

7. Searle, S. R. 1971. *Linear Models*. New York: John Wiley and Sons.

8. Taguchi, G. 1975. *A New Statistical Analysis Method for Clinical Data, the Accumulation Analysis, in Contrast with the Chi-square Test*. Shinjuku Shobo, Tokyo, Japan.

9. Myers, R. N. 1971. *Response Surface Methodology*. Newton, Mass.: Allyn and Bacon.

10. Box, G. E. P., and N. R. Draper. 1969. *Evolutionary Operations: A Statistical Method for Process Improvement*. New York: John Wiley and Sons.

11. Snedecor, G. W., and W. G. Cochran, 1980. *Statistical Methods*. Ames, Iowa: Iowa State University Press.

7

OPTIMIZING THE
WAVE SOLDERING PROCESS

K. M. Lin and R. N. Kackar

7.1 INTRODUCTION

Wave soldering of circuit pack assemblies (CPAs) involves three main phases: fluxing, soldering, and cleaning. The function of each of these three phases is in turn dependent upon a number of factors,[1] and variations in any of these factors can drastically affect the result of the soldering process and CPA reliability. Significant improvement in product quality can be achieved if this process is optimized and all soldered CPAs can go directly into automatic testing without prior inspection and touch-up steps. Moreover, a large amount of direct labor and process cost reductions, as well as appreciable savings, can be realized by reducing in-process inventory and floor-space requirements associated with various inspection, touch-up, and re-testing loops.

A properly chosen water-soluble flux (WSF) can effectively reduce the soldering defect rate and thereby help maximize the solder quality of circuit pack assemblies Cassidy and Lin.[1] In addition to the benefits in cost and quality, a WSF can have advantages in the following areas:

- Soldering of very high-density packages with which a mildly activated rosin flux would have difficulty.

---

1. Glossary equivalent is *control parameter*.

This article originally appeared in the February 1986 issue of *Electronic Packaging and Production* (Cahners Publishing Company). Reprinted with permission.

- Eliminating from the mass soldering areas the use of chlorinated and fluorinated solvents by using aqueous detergent solution to clean the CPAs, thus helping to meet future EPA and OSHA requirements. Solvents are generally expensive to use, and many of them are suspected carcinogens.

- Reducing the electrostatic-discharge damage of very sensitive solid-state devices caused by the brush-cleaning operation with solvents.

- Soldering less solderable components and printed circuit boards (PCBs) is possible due to a higher flux activity.

- Reducing the potential for fire hazard in the plant since different types of flux vehicles are used.

However, due to the aggressive nature of the flux, the post-solder cleaning operation becomes critically important. Any excess amount of flux residue left on the assembly can cause serious problems in product performance and reliability. The cleaning process must, therefore, be closely monitored in order to meet the CPA cleanliness and reliability requirements. For this reason, the designs and materials used for the components and the PCBs must be compatible with the WSF soldering and the aqueous detergent cleaning process. When a rosin type of solderability preservative coating is used on the board, some detergent is added in the solution to enhance the cleaning efficiency. Both the components and the assemblies have to be made irrigable and immersible in the aqueous detergent solution to achieve thorough cleaning and rinsing.

Every factor involved in the entire soldering operation can affect the eventual outcome of the process. Hence, they must be jointly optimized to arrive at a "robust process" from which high-quality products can be manufactured. In a robust process, the operating window of various parameters is made as wide as practical, while the variation in product quality is kept at a minimum. The goal is to achieve a high level of quality the very first time the product is manufactured, without the need for inspection and touch-up operations.

The orthogonal array design method Taguchi[2] was employed in setting up a highly fractionated, designed experiment to optimize the entire fluxing-soldering-cleaning process. In the initial experiment, a total of 17 factors were studied in only 36 test runs. Results from this experiment were then used to set up some smaller scale follow-up experiments to arrive at a robust wave-soldering process which includes an excellent, new flux formulation as well as the proper soldering and cleaning procedures.

7.2 SOLDERING PROCESS

The study presented here illustrates a method for developing robust industrial processes. Further, it indicates that practical considerations in conducting the experiment play a more important role in an industrial environment than the

144

conventional statistical assumptions and models that underlie the design and analysis of experiments.

The WSF soldering process is divided into three phases of operation: fluxing, wave soldering, and aqueous detergent cleaning. Within each phase, there are several factors which can affect the result. These factors and the number of levels of each to be used in this study are listed in Table 7.1.

TABLE 7.1. FACTORS, LEVELS AND THEIR ASSOCIATION WITH OA-TABLE

Factors	No. of levels	Associated columns in Table 7.4
Flux formulation		
A. Type of activator	2	3
B. Amount of activator	3	14
C. Type of surfactant	2	4
D. Amount of surfactant	3	15
E. Amount of antioxidant	3	16
F. Type of solvent	2	5
G. Amount of solvent	3	17
Wave soldering		
H. Amount of flux	3	18
I. Preheat time	3	19
J. Solder temperature	3	13
K. Conveyor speed	3	20
L. Conveyor angle	3	12
M. Wave height setting	2	7
Detergent cleaning		
N. Detergent concentration	2	1
O. Detergent temperature	2	2
P. Cleaning conveyor speed	3	22
Q. Rinse water temperature	2	6

The functions of a soldering flux are to clean the joining surfaces, to protect the cleaned surfaces from re-oxidation, and to lower the surface tension for better solder wetting and solder-joint formation. The types and amounts of activator, antioxidant, surfactant, solvent and vehicle used in a flux mixture all have a definite impact on the soldering efficiency and the cleaning performance of the process. More importantly, these factors can influence the reliability of the product. (More detailed descriptions of the soldering flux can be found in Zado[3]).

Soldering of the assembly is performed in a cascade wave-soldering machine. After flux application and preheating, the non-component side of the assembly is immersed into a solder wave for 1 to 2 seconds, and all solder joints are completed as the CPA exits from the wave. The application method (e.g., foaming, spraying, etc.) and the amount of flux can determine how effectively the joining surfaces are prepared for solder wetting, protected from oxidation, and ultimately, the efficiency of soldering. Large amounts of volatile ingredients in the flux must be driven off by preheating the fluxed assembly to prevent microexplosions from occurring during rapid immersion of the assembly into a molten solder wave. At the same time, gradual preheating can greatly reduce the thermal shock on the assembly during soldering.

Cold solder joints are the result of insufficient heat and temperature during soldering, or premature cooling of the joint after soldering. Very high heat and temperature, on the other hand, can damage the CPA and heat-sensitive components. Conveyor speed can affect the level of preheating and the duration of solder-wave contact, while changes in wave height setting can change the contact length of the board with the solder wave. All these affect the temperature of the joint and the assembly.

The amount of flux and the wave dynamics at the peelback region, where a soldered assembly exits from the solder wave, play important roles in determining the number of solder defects on the wave-soldered CPAs. The position and the shape of the peel-back region can be changed by altering the conveyor angle.

In an aqueous detergent cleaning process, the assembly is first washed with detergent solution, then rinsed with water, and finally dried with hot air jets and heaters. Additional prewash, prerinse, or drying cycles may be added as required. As long as flux residues can be adequately cleaned, the lowest possible detergent concentration should be used so that less detergent is spent and less rinsing is required. The temperature of the detergent solution is raised to achieve more efficient cleaning and also to prevent excess foaming. Similarly, the rinse water is heated to obtain more effective rinsing. Any change in the cleaning conveyor speed changes the assembly dwell time in each cycle of the cleaning process, and that in turn affects the cleanliness of the assembly as well as the production rate of the line.

7.3 EXPERIMENTAL DESIGN

Based on the experience in the flux formulation and soldering/cleaning processes, a decision was made in the beginning to simultaneously study 17 factors in the initial screening experiment as listed in Tables 7.1 and 7.2. Seven of them were at two levels each, and the rest were at three levels each. Test levels of the factors "amount of activator" and "amount of surfactant" depended on the "type of activator" and the "type of surfactant," respectively. Similarly, test levels of "conveyor speed" depended on the associated "solder temperature." The levels for all the factors were deliberately set at extremely wide ranges to have large variations in the response. The trend in response variations thus obtained pointed out a path toward finding a "robust" process.

146

TABLE 7.2. FACTORS AND LEVELS FOR THE EXPERIMENT

Factors	Levels		
	1	2	3
Flux formulation			
A. Type of activator	a	b	—
B. Amount of activator—a	l	m	h
—b	l	m	h
C. Type of surfactant	c	d	—
D. Amount of surfactant—c	l	m	h
—d	l	m	h
E. Amount of antioxidant	l	m	h
F. Type of solvent	x	y	—
G. Amount of solvent	l	m	h
Wave soldering			
H. Amount of flux	l	m	h
I. Preheat time	l	m	h
J. Solder temperature (S.T.)	l	m	h
K. Conveyor speed—l (S.T.)	s	i	f
—m (S.T.)	s	i	f
—h (S.T.)	s	i	f
L. Conveyor angle	l	m	h
M. Wave height setting	l	h	—
Detergent cleaning			
N. Detergent concentration	l	h	—
O. Detergent temperature	l	h	—
P. Cleaning conveyor speed	l	m	h
Q. Rinse water temperature	l	h	—

Note: l: low, m: medium, h: high, s: slow, i: intermediate, f: fast

A soldering flux is a mixture of five ingredients: activator, surfactant, antioxidant, solvent, and vehicle. If the amount of one ingredient is increased, the proportions of other ingredients are decreased accordingly in the mixture. Therefore, in dealing with mixtures, it is necessary to specify the relative amount of each ingredient. One convenient method for this specification (other than specifying percentages of each ingredient) is to write the amount of each ingredient as a fraction of the amount of one particular ingredient. Here, the relative levels were expressed simply as low, medium or high (Table 7.2) to indicate the types of formulations considered.

For 17 factors at their respective 2 and 3 levels, there is a total of 7,558,272 possible combinations. Obviously, it is impractical to vary the experimental factors by the one-at-a-time approach, and some shortcut must be used to carry out such an optimization procedure. In this study, a highly fractionated experimental design was set up by using the orthogonal array design method. Considerations were also given to accommodate some difficult-to-change factors in order to simplify the running of the test.

Table 7.3 shows an orthogonal array, OA (36, 2^{11} x 3^{12}, 2), employed in this design. The array can be used to study up to 11 factors at two levels each and up to 12 factors at three levels each in only 36 test runs. The factors are associated with the columns, and the test levels of the factors are indicated by the numbers in each column. The rows of the OA matrix provide the runs for the experiment. This special design was constructed by combining "A Difference Set" given by Seiden[4] and "A Hadamard Matrix" given by Taguchi.[5]

Because some of the factors involved were more difficult to change, it was necessary in designing the whole experiment to have as few changes as possible for these factors. "Conveyor angle," "detergent concentration," "solder temperature," and "detergent temperature" were such factors. The columns in the OA table, as shown in Table 7.3, were selected for various factors according to the degrees of difficulty in changing them. The levels of the factors associated with columns, 12, 1, 13, and 2 needed to be changed only 2, 5, 8 and 11 times, respectively, when the experiment was conducted according to the row numbers.

The arrangement with which the 17 factors were associated with 17 columns of the orthogonal array (Table 7.3) is shown in Table 7.1. Except for those factors whose test levels require infrequent changes (i.e., associating conveyor angle, detergent concentration, solder temperature and detergent temperature with columns 12, 1, 13, and 2, respectively), the manner in which the other factors were associated with the rest of the columns in OA was arbitrary. Tables 7.4, 7.5, and 7.6 were thus obtained for each of the three phases of the experiment.

Results of the screening experiment were used to design a series of small experiments to confirm the initial findings and to better define the preferable levels for each factor. This series of small experiments consisted of the following:

- *Follow-up experiments* to confirm the findings of the screening experiment and to eliminate any ambiguity.

- *Setup experiments* to better define the levels of significant factors in both the soldering and cleaning processes.

- *Confirmation experiment* to compare the performance of the newly developed flux with a selected commercial flux.

- *Formulation experiment* to fine-tune the composition of the newly developed flux.

- *Large scale trial* to find out the performance of the new flux on a large number of CPAs.

TABLE 7.3. ORTHOGONAL ARRAY TABLE OA (36, $2^{11} \times 3^{12}$, 2)

	N (1)	O (2)	A (3)	C (4)	F (5)	Q (6)	M (7)	— (8)	— (9)	— (10)	— (11)	L (12)	— (13)	B (14)	D (15)	E (16)	G (17)	H (18)	I (19)	K (20)	— (21)	P (22)	— (23)
(1)	1	1	1	1	1	1	1	1	1	1	1	1	1	1	1	1	1	1	1	1	1	1	1
(2)	1	1	1	1	1	1	1	1	1	1	1	1	1	1	1	2	1	2	2	2	1	3	2
(3)	1	1	2	2	2	1	2	2	2	2	2	1	1	2	3	1	3	2	2	1	1	2	1
(4)	1	2	2	2	2	2	2	2	1	1	1	1	2	3	3	1	2	2	1	2	2	3	3
(5)	1	2	1	1	1	2	1	1	1	1	1	1	2	3	2	3	1	3	2	3	3	2	1
(6)	1	2	2	2	2	1	2	1	1	1	1	1	2	3	1	2	3	1	3	2	2	3	3
(7)	1	1	2	1	1	2	1	2	2	2	1	1	2	1	1	1	3	3	3	1	1	1	3
(8)	2	1	2	2	2	2	2	1	1	1	1	1	3	1	2	2	1	1	2	3	3	1	2
(9)	2	1	1	1	1	2	1	2	2	2	2	1	3	2	2	1	2	3	1	1	2	1	2
(10)	2	2	2	2	2	2	1	1	1	1	1	1	3	2	3	3	2	2	3	2	1	3	3
(11)	2	2	2	1	1	2	1	2	2	2	2	1	3	3	1	3	1	1	1	3	1	3	1
(12)	2	2	1	2	2	1	2	2	2	2	1	1	3	3	2	1	2	3	2	2	1	1	2
(13)	1	1	2	1	1	2	1	1	1	1	1	2	1	2	3	3	2	1	3	3	2	3	3
(14)	1	1	1	2	2	2	2	2	2	2	1	2	1	2	1	2	3	2	1	1	3	1	1
(15)	1	1	1	1	1	1	1	1	1	1	2	2	1	3	2	1	1	3	2	2	3	2	3
(16)	1	2	1	2	2	2	2	1	1	1	1	2	2	3	2	3	3	1	2	1	1	3	3
(17)	2	2	2	1	1	1	1	2	2	2	1	2	2	1	3	1	2	3	3	3	3	2	1
(18)	2	2	2	2	2	2	2	2	1	1	1	2	2	1	1	3	1	2	1	2	2	1	2
(19)	2	1	1	1	1	1	2	1	1	1	1	2	2	2	3	3	3	2	1	3	3	2	2
(20)	2	1	2	2	2	1	2	2	2	2	1	2	2	2	1	1	1	3	2	1	1	3	3
(21)	2	1	2	1	1	1	1	1	1	1	2	2	3	3	2	2	2	1	3	2	2	2	1
(22)	1	2	2	2	2	2	2	2	2	2	2	2	3	1	3	1	3	2	1	1	1	1	2
(23)	1	2	1	1	1	1	1	2	2	2	2	2	3	2	1	3	2	3	2	2	2	3	3
(24)	1	2	1	2	2	2	2	1	1	1	1	2	3	2	2	2	1	1	3	3	3	2	1
(25)	1	1	2	1	1	2	1	2	2	2	2	3	1	1	3	2	3	3	2	1	3	2	2
(26)	2	1	2	2	2	2	2	2	2	2	1	3	1	2	1	3	1	1	3	2	1	2	3
(27)	2	1	1	1	1	2	1	1	1	1	2	3	1	3	2	1	2	2	1	3	3	3	1
(28)	2	2	2	2	2	1	2	1	2	2	1	3	2	2	3	3	3	1	2	2	3	3	3
(29)	2	2	2	1	1	1	1	2	1	1	2	3	2	3	1	2	1	3	3	3	1	2	2
(30)	1	2	1	2	2	1	2	2	1	1	2	3	2	1	2	3	2	2	1	1	2	3	1
(31)	2	2	2	2	1	2	1	1	2	2	2	3	3	1	3	1	2	2	2	3	1	2	1
(32)	2	1	1	1	2	2	2	2	1	1	2	3	3	2	1	3	2	3	3	1	2	3	3
(33)	2	2	2	2	1	1	1	1	2	2	1	3	3	3	2	2	3	1	1	2	1	1	2
(34)	2	2	1	1	1	2	2	2	2	2	1	3	1	1	2	3	1	2	3	3	3	2	1
(35)	2	2	2	2	2	1	1	1	2	2	2	3	2	2	3	2	2	3	1	1	1	3	2
(36)	2	2	1	1	1	1	2	1	2	2	1	3	2	3	1	1	1	3	3	3	3	1	2

TABLE 7.4. RUNS FOR FLUX FORMULATION

Flux formulation	Type of activator	Amount of activator	Type of surfactant	Amount of surfactant	Amount of antioxidant	Type of solvent	Amount of solvent
(1)	a	al	c	cm	m	x	l
(2)	a	al	c	cl	h	x	l
(3)	b	bm	d	dl	l	y	h
(4)	a	ah	d	dh	l	y	m
(5)	b	bh	c	ch	l	y	i
(6)	b	bh	d	dm	h	x	m
(7)	b	bl	d	dl	h	x	m
(8)	b	bm	c	ch	m	y	h
(9)	a	am	d	dh	m	y	l
(10)	b	bm	c	cl	l	x	m
(11)	a	ah	d	dm	h	y	h
(12)	a	al	c	cm	m	x	m
(13)	a	am	c	ch	h	x	m
(14)	a	am	c	cm	m	y	m
(15)	b	bh	d	dm	m	y	l
(16)	a	al	d	dl	m	y	h
(17)	b	bl	c	cl	m	y	m
(18)	b	bl	d	dh	l	x	h
(19)	b	bm	c	dm	l	x	l
(20)	b	bh	c	cl	h	y	m
(21)	a	ah	d	dl	h	y	h
(22)	b	bh	c	cm	m	x	m
(23)	a	al	d	dh	l	x	l
(24)	a	am	c	ch	h	y	h
(25)	a	ah	c	cl	l	x	m
(26)	a	ah	c	ch	m	x	h
(27)	b	bl	d	dh	h	y	l
(28)	a	am	d	dm	h	y	h
(29)	b	bm	c	cm	m	x	m
(30)	b	bm	d	dl	m	x	l
(31)	b	bh	c	dh	l	y	h
(32)	b	bl	c	cm	l	y	l
(33)	a	al	d	ch	h	y	h
(34)	b	bl	c	dl	m	x	l
(35)	a	am	d	dl	m	x	m
(36)	a	bh	c	cl	l	y	l

TABLE 7.5. RUNS FOR WAVE SOLDERING

Soldering	Conveyor angle (degree)	Solder temperature (F)	Conveyor speed (feet/min)	Wave height (setting)	Preheat time (seconds)	Amount of flux (grams)
(1)	l	l	ls	l	m	l
(2)	l	l	lf	h	l	h
(3)	l	l	ls	l	h	m
(4)	l	l	li	h	l	l
(5)	l	m	mf	l	m	m
(6)	l	m	mf	h	h	h
(7)	l	m	mi	h	h	l
(8)	l	m	ms	h	l	h
(9)	l	h	hf	l	h	l
(10)	l	h	hi	l	m	h
(11)	l	h	hs	l	m	m
(12)	l	h	hi	h	l	m
(13)	m	m	mi	l	h	m
(14)	m	m	ms	h	m	l
(15)	m	m	mi	l	l	h
(16)	m	m	mf	h	m	m
(17)	m	h	hs	l	h	h
(18)	m	h	hs	h	l	l
(19)	m	h	hf	h	l	m
(20)	m	h	hi	h	m	l
(21)	m	l	ls	l	l	m
(22)	m	l	lf	l	h	l
(23)	m	l	li	l	h	h
(24)	m	l	lf	h	m	h
(25)	h	h	hf	l	l	h
(26)	h	h	hi	h	h	m
(27)	h	h	hf	l	m	l
(28)	h	h	hs	h	h	h
(29)	h	l	li	l	l	l
(30)	h	l	li	h	m	m
(31)	h	l	ls	h	m	h
(32)	h	l	lf	h	h	m
(33)	h	m	mi	l	m	h
(34)	h	m	ms	l	l	m
(35)	h	m	mf	l	l	l
(36)	h	m	ms	h	h	l

TABLE 7.6. RUNS FOR DETERGENT CLEANING

Cleaning	Detergent concentration (percent)	Detergent temperature (F)	Cleaning conveyor speed (feet/min.)	Rinse water temperature (F)
(1)	l	l	h	l
(2)	l	l	l	h
(3)	l	l	m	l
(4)	l	h	l	l
(5)	l	h	h	h
(6)	l	h	m	h
(7)	h	l	h	l
(8)	h	l	l	l
(9)	h	l	m	l
(10)	h	h	h	l
(11)	h	h	l	h
(12)	h	h	m	l
(13)	l	l	l	l
(14)	l	l	m	h
(15)	l	l	h	l
(16)	l	h	m	l
(17)	l	h	l	h
(18)	l	h	h	h
(19)	h	l	l	l
(20)	h	l	m	h
(21)	h	l	h	h
(22)	h	h	l	l
(23)	h	h	m	h
(24)	h	h	h	l
(25)	l	l	m	l
(26)	l	l	h	h
(27)	l	l	l	l
(28)	l	h	h	l
(29)	l	h	m	h
(30)	l	h	l	h
(31)	h	l	m	l
(32)	h	l	h	h
(33)	h	l	l	h
(34)	h	h	m	l
(35)	h	h	h	h
(36)	h	h	l	l

7.4 RESPONSES

The aim of the experiment was to determine those factor levels which would result in the least solder defects, the smallest amount of flux residues, the highest insulation resistance (IR) for the IR test coupons, and the least amount of solder-mask cracking. Normally, solder defects can be classified into four categories: no solder, insufficient solder, crosses, and icicles. However, the analysis of initial soldering data indicated that almost all defects were either crosses or icicles. Therefore, only the crosses and icicles were investigated.

The amount of flux residue left on the assembly could affect the performance and reliability of the product. Therefore, chemical analyses were conducted to identify the quantities of residues on the assembly after it had been processed through the detergent cleaner.

The electrical performance of the processed test board was evaluated by the insulation resistance test of comb-patterned coupons in an environmental chamber. For each coupon, four IR values were taken: at initial ambient condition, after 30 minutes, 1 day, and 4 days in the chamber. Two chamber conditions were used; one was at a normal stress level (35 C, 90 percent relative humidity, and no bias voltage), and the other was at a high stress level (65 C, 90 percent relative humidity, and no bias voltage). The higher the insulation resistance reading, the better the electrical performance.

If the solder-mask material on the circuit board were incompatible with either the flux or the process, damages or cracks could occur in the solder mask. This could result in the corrosion or staining of the circuit. After completing the IR test, all comb-patterned coupons were visually inspected, and the numbers of stained spots were recorded.

The responses examined for this experiment are summarized in Table 7.7. The final performance of a particular fluxing-soldering-cleaning process was evaluated on the basis of both the soldering defect rate and the board reliability as measured by the IR test result.

TABLE 7.7. RESPONSES

Insulation resistance test	Cleaning characterization
Measured at initial room condition, 30 min, 1 and 4 days under:	Measure the amount of residues left on the board
- 35C, 90 percent RN, no bias voltage - 65 C, 90 percent RH, no bias voltage	
Soldering efficiency	**Solder mask cracking**
Visual inspection to record the number of no solder, insufficient solder, good solder, excess solder, and miscellaneous defects.	Visual inspection to record the number of cracked spots on the solder mask of the IR coupons

7.5 EXPERIMENTS AND RESULTS

This experiment used a special test board fashioned after an actual product and very similar to the test board used in an earlier study Cassidy and Lin.[1] The board was a double sided, rigid board with Sealbrite (a registered trademark of London Chemical Co.) copper circuitry and plated through holes. The components selected were historically difficult to solder, maximizing the number of soldering defects on each board to produce optimal statistics with the smallest number of boards. Solder mask was used on both sides of the board except on the copper land and certain special circuit pattern areas.

Every board was clearly identified in six places so that various portions of the board, after sectioning for different tests and analysis, could later be traced for re-examination or verification.

Thirty-six different water-soluble fluxes were formulated according to the experimental design. Two boards were used in each run under the conditions listed in Table 7.2. Test boards were soldered in each run in a cascade wave-soldering machine and cleaned in an aqueous detergent spray cleaner, under the conditions listed in Table 7.2. Then the boards were examined for soldering defects, sectioned, and analyzed for residues. Because of the capacity and the availability of the environmental testing chamber, comb-patterned coupons were accumulated in separate plastic bags at room conditions before they were placed into the chamber at the same time to facilitate the IR test. IR values of the coupons were recorded under the specified test conditions as mentioned before.

One of the 36 fluxes was used in the subsequent "soldering follow-up experiment," and four new fluxes with slight composition variations were used in the "flux formulation follow-up experiment." From the available information up to that point, a new flux "A" was formulated and tested in the setup experiments to gain soldering performance information as well as to better define the factor levels in both the soldering and cleaning processes. A new flux "B" was then developed for the confirmation experiment, and the soldering performance of the new flux B was evaluated against a selected commercial flux. Two soldering conditions with different solder temperature, top-side board temperature after preheat, and conveyor speed, were used for this experiment.

Condition 1 was commonly used in a production line, while condition 2 was determined from the designed experiment for the newly developed flux B. Two major types of soldering defects were found—joint crosses and fineline crosses. Table 7.8 shows the total number of solder defects for each flux used under both soldering conditions (10 boards each); a bar graph for this result is shown in Figure 7.1.

Analysis of the result showed that there was a significant performance improvement by the use of flux B under either soldering condition. Furthermore, there was a smaller 95 percent confidence interval for flux B in both cases. It indicated that soldering with flux B gave a smaller variation and was a very robust process.

TABLE 7.8. NORMALIZED SOLDERING DEFECTS PER BOARD*

Flux	Mean joint crosses	95 percent confidence interval	Mean fine-line crosses	Total joint crosses	Soldering condition
Commercial	0.92	0.45	0.08	1.00	1
Flux B	0.54	0.22	0	0.54	
Commercial	0.65	0.28	0.23	0.88	
Flux B	0.44	0.16	0.06	0.50	2

* Normalized with the total number of defects for the commercial flux soldered under condition 1.

Results for both the insulation resistance test and the cleaning characterization were all very good for flux B; they passed all the required specifications. Visual inspections performed on the solder mask indicated that the smallest number of cracks occurred with the lowest soldering temperature.

The orthogonal array design method employed in designing the experiment is a very powerful technique by which a complex process with a large number of factors can be systematically and effectively studied with a series of small experiments, while practical considerations in industrial experimentation have been taken into account. In addition to being able to identify the significant factors, it can also provide information on the relative importance of various interactions. This method has been successfully applied in various industries in Japan to upgrade the product and lower the production cost.

The concept of such an off-line quality control method can be used both in the early stages of product designs and in the manufacturing processes to achieve a complete system's optimization and to identify a process which is relatively insensitive to the small variations in the process parameters. The high level of product quality is achieved by doing the job right the first time rather than trying to obtain it through corrective actions after the product is made. In most cases, both lower cost and high quality of a product can be realized at the same time Crosby.[6]

The experiment discussed is an ambitious undertaking. A very lean design with only 36 test runs is used in the initial screening experiment. The process restrictions have also been taken into consideration so that a minimum number of changes are required of those difficult-to-change factors. The levels of each factor have been deliberately set at extreme conditions in order to obtain a wide range of responses which will in turn point out the path toward finding a "robust" process. In the subsequent series of small experiments, more definite conditions and factor levels have been obtained, and a superior WSF, along with its appropriate soldering-cleaning processes, has been developed. With some minor adjustments in the composition for a flux formulation experiment, the newly developed WSF can be further fine-tuned to arrive at an optimum flux formulation for the circuit pack wave-soldering process.

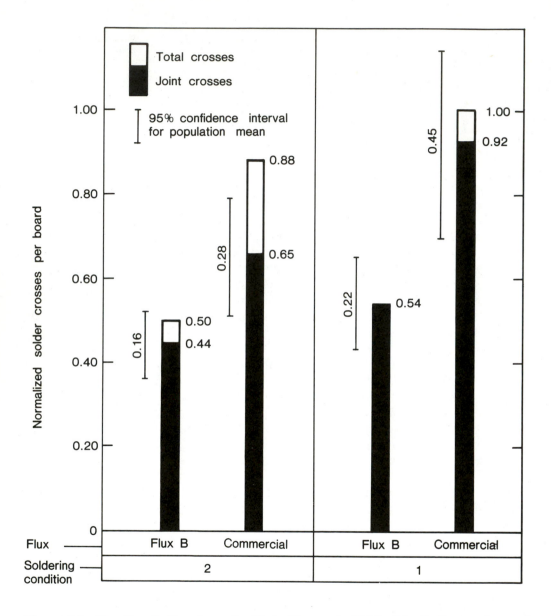

Figure 7.1. Bar Graph Based on the Data Shown in Table 7.8. Indicates the total number of solder defects for each flux used under both soldering conditions.

ACKNOWLEDGMENTS

This study is the result of a combined effort by a large group of people from the solder process department of the AT&T Engineering Research Center and the Quality Assurance Center of the AT&T Bell Laboratories at Holmdel, NJ. The authors wish to acknowledge the valuable contribution made by each of them. They also wish to express their special thanks to F.M. Zado who, with his in-depth knowledge of the flux formulations, actively participated in this undertaking.

REFERENCES

1. Cassidy, M. P., and K. M. Lin. 1981. Soldering Performance of Fluxes: A Study. *Electronic Packaging and Production* **21** (Nov. no. 11):153-164.

2. Taguchi, Genichi. 1981. *An Introductory Text for Design Engineers: Design and Design of Experiments*. Tokyo: Japanese Standards Association.

3. Zado, F. M. 1981. The Theory and Practice of High Efficiency Soldering with Noncorrosive Rosin Soldering Fluxes. *Proceedings of the Printed Circuit World Convention II.* Vol. I. Munich, West Germany. June.

4. Seiden, E. 1954. On the Problem of Construction of Orthogonal Arrays. *Annals of Mathematical Statistics* **25**:151-156.

5. Taguchi, Genichi. 1980. *Introduction to Off-Line Quality Control*. Tokyo: Japanese Standards Association.

6. Crosby, P. B. 1979. *Quality is Free*. New York: McGraw-Hill.

8

ROBUST DESIGN:
A COST-EFFECTIVE METHOD FOR
IMPROVING MANUFACTURING PROCESSES

Raghu N. Kackar and
Anne C. Shoemaker

8.1 THE CENTRAL IDEA

A main cause of poor yield in manufacturing processes is manufacturing variation. These manufacturing variations include variation in temperature or concentration within a production batch, variation in raw materials, and drift of process parameters. The more sensitive a process is to these variations, the more expensive it is to control.

The method described in this paper, robust design, is a cost-effective approach to improve yield. It uses statistically planned experiments to identify process control parameter settings that reduce the process' sensitivity to manufacturing variation. This method is based on the technique of robust design developed by Professor Genichi Taguchi, a Japanese expert on quality.

What is the central idea of robust design? The easiest way to explain it is to look at how robust design is improving the process for making optical filters. These filters, which are used in wavelength-division multiplexers, consist of a quartz substrate coated with thin layers of titanium dioxide and silicon dioxide.

Variability of the filter's index of refraction in manufacture and in the field is a major problem, and one cause is the change in relative humidity in the factory and under field-use conditions. Figure 8.1 shows a simplified filter cross section with only one layer of titanium dioxide. Water molecules from the environment condense into pores in the film and change the filter's overall index of refraction.

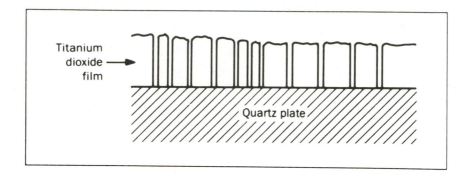

Figure 8.1. Cross Section of a Simple Filter With Only One Film Layer on a Quartz Substrate

To solve this problem, we could keep the relative humidity constant by building an hermetic seal around the filter. But this is a difficult and expensive solution. Instead, we are taking another approach: Design the filter-making process so that the film is dense with few crevices to trap water molecules. Thus, the filter's index of refraction will be less sensitive to humidity changes.

This idea—make a product or process insensitive to variation—is the essence of robust design. However, this paper concentrates on manufacturing process design. Although most ideas presented here also apply to product design, other optimization methods may be more appropriate.

In this paper, we formulate the robust-design problem and give four operational steps for applying the method. Then, we show how we used the method to improve an integrated-circuit (IC) fabrication process. Finally, we discuss the relationship between robust design and traditional uses of statistical design of experiments.

8.2 FORMULATING THE ROBUST DESIGN PROBLEM

To define the objective of robust design more precisely, we need three concepts: functional characteristics, control parameters, and sources of noise.

Functional characteristics are basic, measurable quantities that determine how well the final product functions. For the optical-filter example, the film's index of refraction and absorption are two of the characteristics that determine how well the completed filter separates light of certain wavelengths.

When we study a specific step of a manufacturing process, the functional characteristics are usually measurements that can be made on the incomplete product soon after that step.

The essence of robust design is to reduce variation of a product's or process's functional characteristics. (The average value of a functional characteristic is usually of secondary concern, because we can adjust it to the target after we minimize variability. In a later section, we will see how this was possible in the manufacture of IC wafers.) Two types of variables affect functional characteristics: control parameters and sources of noise.

Control parameters are the controllable process variables; their operating standards can be specified by the process engineers. In the optical-filter example, the control parameters include the temperature of the substrates and the method of cleaning the substrates.

In contrast, *sources of noise* are the variables that are impossible or expensive to control. Examples include temperature and humidity variations in the factory, drift in process control parameters, and variation in raw materials. They, in turn, cause variations in the product's functional characteristics.

The objective in robust design is to find those control parameter settings where noise has a minimal effect on the functional characteristics. As in the optical-filter example, the key idea is to reduce functional characteristic sensitivity by making the process insensitive to noise rather than by controlling the sources of noise.

To attain this objective, we systematically vary the control parameters in an experiment and measure the effect of noise for each experimental run. Then we use the results to predict which control parameter settings will make the process insensitive to noise.

8.3 OPERATIONAL STEPS FOR ROBUST DESIGN

The robust design method can be applied by following four operational steps:

1. List the functional characteristics, control parameters, and sources of noise.

2. Plan the experiment:

    a. How will control parameter settings be varied?

    b. How will the effect of noise be measured?

3. Run the experiment and use the results to predict improved control parameter settings.

4. Run a confirmation experiment to check the prediction.

The next section describes each step and how it was applied to improve the process of growing an epitaxial layer on silicon wafers used in IC fabrication.

8.4 EPITAXIAL-PROCESS EXAMPLE

A first step in processing silicon wafers for IC devices is to grow an epitaxial layer on polished silicon wafers. As Figure 8.2 shows, a batch of 14 wafers is processed together. Two wafers are mounted on each side, or facet, of a seven-sided cylindrical structure, called a susceptor.

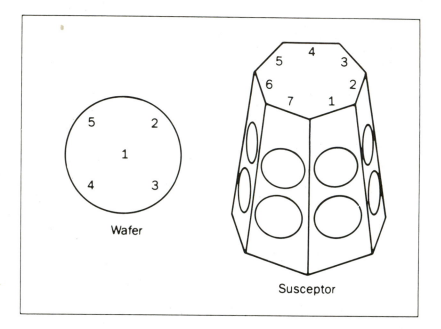

Figure 8.2. Plan for Studying the Effect of Noise on the Epitaxial Process. Five measurements of epitaxial thickness are made on each of 14 wafers.

The susceptor is positioned inside a metal bell jar. As the susceptor rotates, chemical vapors are introduced through nozzles near the top of the jar. The susceptor and wafers are heated to an elevated temperature, which is maintained until the epitaxial layer is thick enough.

The process specifications called for a layer between 14 and 15 micrometers thick, but the actual variation from the ideal 14.5 micrometers was greater than this. Therefore, we needed to minimize nonuniformity in the epitaxial layer, yet keep the average thickness as close to the ideal as possible.

While working with the engineers responsible for the process, we conducted a robust design experiment to reduce the variability, but still keep the average thickness near 14.5 micrometers. We identified eight key process-control parameters and, using 16 batches of wafers, evaluated the epitaxial layer's average thickness and nonuniformity for two test settings of each parameter.

The experiment's results showed that two control parameters, nozzle position and susceptor-rotation method, determined the epitaxial layer's uniformity. The results also showed that one parameter, deposition time, had a large effect on average thickness but no effect on uniformity. A follow-up experiment confirmed that the new settings for nozzle position and rotation method, determined in the first experiment, reduced nonuniformity about 60 percent. The change to new settings did not increase cost.

The next subsections describe how the four operational steps of robust design were followed to obtain these new parameter settings.

8.4.1 STEP 1—LIST FUNCTIONAL CHARACTERISTICS, CONTROL PARAMETERS, AND SOURCES OF NOISE

Most processes have many functional characteristics. For example, the epitaxial process has two important functional characteristics: epitaxial thickness and epitaxial resistivity. Both were studied, but this paper will discuss only the results for epitaxial thickness.

Although every process has many control parameters, we usually cannot study them all simultaneously. Most robust design experiments study five to ten control parameters at a time.

In the epitaxial-process experiment, we studied eight parameters; A through H represent the code letters used in this paper for these parameters:

A. Susceptor-rotation method

B. Wafer code

C. Deposition temperature

D. Deposition time

E. Arsenic gas flow rate

F. Hydrochloric acid etch temperature

G. Hydrochloric acid flow rate

H. Nozzle position.

Parameter B, wafer code, is not a control parameter in the usual sense. Many different codes of wafers must pass through the epitaxial process. However, including the code as a control parameter allowed us to identify the settings of other control parameters that produced uniform epitaxial layers on a variety of wafer codes.

We tested each parameter at two settings. Table 8.1 shows the initial settings and the two test settings.

TABLE 8.1. INITIAL AND TEST SETTINGS OF EACH CONTROL PARAMETER

Control Parameter	Initial Setting	Test Setting 0	Test Setting 1
A. Susceptor-rotation method	Oscillating	Continuous	Oscillating
B. Code of wafers	--	668G4	678D4
C. Deposition temperature	1215°C	1210°C	1220°C
D. Deposition time	Low	High	Low
E. Arsenic flow rate	57%	55%	59%
F. Hydrochloric acid etch temperature	1200°C	1180°C	1215°C
G. Hydrochloric acid flow rate	12%	10%	14%
H. Nozzle position	4	2	6

The principal sources of noise in the epitaxial process were uneven temperature, vapor-concentration, and vapor-composition profiles inside the bell jar. Because of these uneven profiles, the epitaxial-layer thickness was different on wafers at different locations on the susceptor. The difference was largest between wafers at the top and bottom positions on each facet.

8.4.2 STEP 2—PLAN THE EXPERIMENT

In a robust design experiment, we vary the settings of the control parameters simultaneously in a few experimental runs. For each run, we make multiple measurements of the functional characteristic to evaluate the process's sensitivity to noise.

Therefore, planning the experiment is a two-part step that involves deciding how to vary the parameter settings and how to measure the effect of noise.

Step 2a—Plan How Control Parameter Settings Will Be Varied. In step 1, we identified eight control parameters as potentially important in the epitaxial process. Because we would test each parameter at two settings, we would need 2^8, or 256, experimental runs to evaluate every possible combination of settings. Clearly, this would be too expensive and time consuming. Fortunately, we could choose a small subset of these runs and still obtain the most important information about the control parameters.

In the epitaxial-process example, we conducted only 16 of the 256 possible runs. Table 8.2, called the *control array* for the experiment, shows the control-parameter settings for the 16 runs.

TABLE 8.2. CONTROL ARRAY FOR EPITAXIAL PROCESS EXPERIMENT

Experimental Run	Control Parameter							
	A	B	C	D	E	F	G	H
1	Cont	668G4	1210	High	55	1180	10	2
2	Cont	668G4	1210	High	59	1215	14	6
3	Cont	668G4	1220	Low	55	1180	14	6
4	Cont	668G4	1220	Low	59	1215	10	2
5	Cont	678D4	1210	Low	55	1215	10	6
6	Cont	678D4	1210	Low	59	1180	14	2
7	Cont	678D4	1220	High	55	1215	14	2
8	Cont	678D4	1220	High	59	1180	10	6
9	Osclt	668G4	1210	Low	55	1215	14	2
10	Osclt	668G4	1210	Low	59	1180	10	6
11	Osclt	668G4	1220	High	55	1215	10	6
12	Osclt	668G4	1220	High	59	1180	14	2
13	Osclt	678D4	1210	High	55	1180	14	6
14	Osclt	678D4	1210	High	59	1215	10	2
15	Osclt	678D4	1220	Low	55	1180	10	2
16	Osclt	678D4	1220	Low	59	1215	14	6

NOTE: The control array specifies the experimental runs.

This control array has an important property: it is *balanced*. That is, for every pair of parameters, each combination of test settings appears an equal number of times. For example, consider the parameters: deposition temperature (C) and deposition time (D). The combinations (C = 1210°C, D = High), (C = 1210°C, D = Low), (C = 1220°C, D = High), and (C = 1220°C, D = Low) each occur four times in columns 3 and 4.

Because of this balancing property, it is meaningful to compare the two deposition-temperature test settings over a range of test settings for deposition time and each of the other control parameters in the experiment. Therefore, we can draw separate conclusions about each control parameter from the results of the experiment.

By contrast, it can be difficult to draw conclusions from the results of an unbalanced experiment. For example, a "one-parameter-at-a-time" experiment, the type most commonly run, is unbalanced. In such experiments, one control parameter is varied at a time, while all others are held fixed. Table 8.3 shows this unbalanced experiment for the epitaxial-process example.

If we took this approach, we would compare deposition-temperature settings (C) using the results of runs 3 and 4. However, the comparison would be valid *only* when the other seven control parameters are fixed at their values in these two runs.

TABLE 8.3. A ONE-PARAMETER-AT-A-TIME EXPERIMENT

Experimental	Control Parameter							
Run	A	B	C	D	E	F	G	H
1	Cont	668G4	1210	High	55	1180	10	2
2	Osclt	668G4	1210	High	55	1180	10	2
3	Osclt	678D4	1210	High	55	1180	10	2
4	Osclt	678D4	1220	High	55	1180	10	2
5	Osclt	678D4	1220	Low	55	1180	10	2
6	Osclt	678D4	1220	Low	59	1180	10	2
7	Osclt	678D4	1220	Low	59	1215	10	2
8	Osclt	678D4	1220	Low	59	1215	14	2
9	Osclt	678D4	1220	Low	59	1215	14	6

NOTE: This is not a balanced experiment.

This example also illustrates a second advantage of balanced experiments. In the balanced epitaxial-layer experiment (Table 8.2), we use the results of all 16 runs to compare deposition-temperature settings (C). This is more precise than the one-parameter-at-a-time comparison, which uses only two runs.

Control arrays that have the desirable balancing property can be constructed easily from special tables called orthogonal arrays (OAs). We constructed the control array for the epitaxial-process experiment from the orthogonal array, OA_{16} (Table 8.4).

To obtain the control array, we assigned the control parameters to columns 1, 2, 4, 7, 8, 11, 13, and 14 of OA_{16}, and replaced the symbols 0 and 1 with each parameter's test settings. This particular assignment of parameters to the columns of OA_{16} produces a *resolution-IV plan*; it ensures that the experiment will give good estimates of each control parameter's first-order (linear) effects. (For more on the resolution of experiment plans, see Box, Hunter, and Hunter, Chapter 12[1].)

Although OA_{16} is one of the most frequently used orthogonal arrays, there are many other useful arrays. A particularly valuable one is OA_8, which permits studying up to four control parameters at two levels each in only eight experimental runs. Arrays such as OA_{18} and OA_{27} are useful when most control parameters have three test settings. Phadke and others[2] used OA_{18} to construct the control array for an experiment to improve a window-photolithography process.

Step 2b—Plan How The Effect Of Noise Will Be Measured. In the epitaxial-layer experiment, a major cause of variation in epitaxial thickness was uneven distribution of chemical vapors from the top to the bottom of the bell jar. To reflect this variation, we systematically measured the epitaxial thickness of wafers at top and bottom locations on each susceptor facet.

Our plan was to measure epitaxial thickness at five places (Figure 8.2) on each of the 14 wafers, a total of 70 measurements from one run. We did this for all 16 experimental runs in the control array.

TABLE 8.4. ORTHOGONAL ARRAY OA_{16} (2^{15})

Experimental Run	A	B		C			D	E			F		G	H	
	1	2	3	4	5	6	7	8	9	10	11	12	13	14	15
1	0	0	0	0	0	0	0	0	0	0	0	0	0	0	0
2	0	0	0	0	0	0	0	1	1	1	1	1	1	1	1
3	0	0	0	1	1	1	1	0	0	0	0	1	1	1	1
4	0	0	0	1	1	1	1	1	1	1	1	0	0	0	0
5	0	1	1	0	0	1	1	0	0	1	1	0	0	1	1
6	0	1	1	0	0	1	1	1	1	0	0	1	1	0	0
7	0	1	1	1	1	0	0	0	0	1	1	1	1	0	0
8	0	1	1	1	1	0	0	1	1	0	0	0	0	1	1
9	1	0	1	0	1	0	1	0	1	0	1	0	1	0	1
10	1	0	1	0	1	0	1	1	0	1	0	1	0	1	0
11	1	0	1	1	0	1	0	0	0	1	0	1	1	0	0
12	1	0	1	1	1	0	1	0	1	0	1	0	1	0	1
13	1	1	0	0	1	1	0	0	1	1	0	0	1	1	0
14	1	1	0	0	1	1	0	1	0	0	1	1	0	0	1
15	1	1	0	1	0	0	1	0	1	1	0	1	0	0	1
16	1	1	0	1	0	0	1	1	0	0	1	0	1	1	0

NOTE: This assignment of control parameters A through H to the columns of OA_{16} provides a resolution-IV plan.

This plan for measuring the effect of noise is typical of procedures followed in most robust-design experiments for process improvement. Usually, we measure the effect of noise by taking multiple measurements of the functional characteristic at different positions on a unit and on several units in the experimental run. It is important to plan these measurements to reflect the effect of noise and follow the same plan for each experimental run, so comparisons between the runs are fair.

8.4.3 STEP 3—RUN THE EXPERIMENT AND USE THE RESULTS TO PREDICT IMPROVED CONTROL PARAMETER SETTINGS

If the experiment is well-planned and run according to plan, the analysis needed to predict improved parameter settings is simple.

As described in step 2, we made 70 measurements of epitaxial thickness for each of the 16 test runs in Table 8.2. For each run, we calculated the mean and variance of these measurements. If y_1 through y_{70} represent the 70 measurements for one test run, the mean is

$$\bar{y} = \frac{1}{70} \sum_{i=1}^{70} y_i \, ,$$

and the variance is

$$s^2 = \frac{1}{69} \sum_{i=1}^{70} (y_i - \bar{y})^2 \, .$$

Table 8.5 shows the values of $\bar{y}$ and $\log s^2$ for each test run. (The logarithm of s^2 is taken to improve statistical properties of the analysis.)

TABLE 8.5. MEAN AND LOG OF VARIANCE OF EPITAXIAL THICKNESS FOR EACH TEST RUN

Experimental Run	Mean $\bar{y}$ (μm)	Log of Variance $\log s^2$
1	14.821	−0.4425
2	14.888	−1.1989
3	14.037	−1.4307
4	13.880	−0.6505
5	14.165	−1.4230
6	13.860	−0.4969
7	14.757	−0.3267
8	14.921	−0.6270
9	13.972	−0.3467
10	14.032	−0.8563
11	14.843	−0.4369
12	14.415	−0.3131
13	14.878	−0.6154
14	14.932	−0.2292
15	13.907	−0.1190
16	13.914	−0.8625

    The experiment's objective was to make the epitaxial-thickness variance small and make the mean thickness close to 14.5 micrometers. We focused on reducing the variance of epitaxial thickness. After minimizing variance, we could adjust deposition time to get a 14.5-micrometer mean thickness.

    To reduce the variance of epitaxial thickness, we identified the control parameters that have the largest effect on $\log s^2$ and set them at the settings that minimized $\log s^2$. Table 8.6 and Figure 8.3 display the average values of $\log s^2$ for the test settings of all eight control parameters.

TABLE 8.6. AVERAGE VALUE OF $\log s^2$ AT DIFFERENT TEST SETTINGS

Control Parameter	Average $\log s^2$		
	Test Setting 0	Test Setting 1	Difference
A. Susceptor-rotation method	−0.8245	−0.4724	0.3521
B. Code of wafers	−0.7095	−0.5875	0.1220
C. Deposition temperature	−0.7011	−0.5958	0.1053
D. Deposition time	−0.5237	−0.7732	−0.2495
E. Arsenic gas flow rate	−0.6426	−0.6543	−0.0117
F. Hydrochloric acid etch temperature	−0.6126	−0.6843	−0.0717
G. Hydrochloric acid flow rate	−0.5980	−0.6989	−0.1008
H. Nozzle position	−0.3656	−0.9313	−0.5658

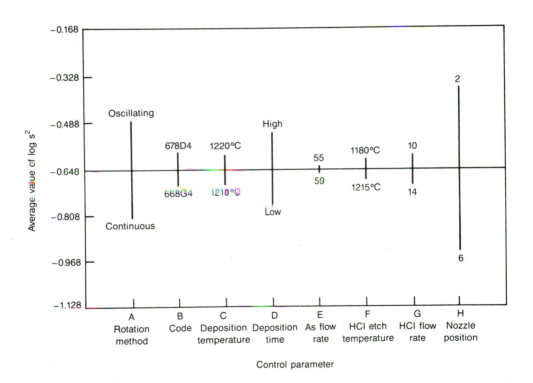

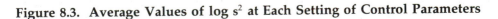

Figure 8.3. Average Values of $\log s^2$ at Each Setting of Control Parameters

169

For example, consider design parameter A, susceptor-rotation method. Its setting is *continuous* in runs 1 through 8 and *oscillating* in runs 9 through 16.

From Table 8.6, the average of the values of $\log s^2$ for runs 1 through 8 is

$$-0.8245 = \frac{-0.4425 + -1.1989 + \cdots + -0.6270}{8}.$$

Likewise, the average of the values of $\log s^2$ for runs 9 through 16 is

$$-0.4724 = \frac{-0.3467 + -0.8563 + \cdots + -0.8625}{8}.$$

Because $\log s^2$ measures the uniformity of epitaxial thickness, the better setting of each control parameter is the one that gives the smaller average value of $\log s^2$. For example, continuous rotation (setting 0) of the susceptor is better than oscillation (setting 1) because, on the average, it gives more uniform epitaxial layers.

We compared the test settings of the other control parameters in the same way. However, we changed each parameter's setting only if this change had a large effect on the variability of epitaxial thickness. When a control parameter is tested at two settings, the magnitude of its effect on variability is measured by the difference between the average values of $\log s^2$ at those settings.

The lengths of the bars in Figure 8.3 provide a visual summary of the magnitude of each control parameter's effect. (The horizontal line at -0.648 is the average value of $\log s^2$ over all 16 runs.) Because nozzle position and rotation method had greater effect than the other six control parameters, we changed the settings of only these two parameters. All others were left at their initial settings.

8.4.4 STEP 4—RUN A CONFIRMATION EXPERIMENT TO CHECK THE PREDICTION

When we used OA_{16} to construct the control array for the epitaxial-process experiment, we assumed that the relationship between $\log s^2$ and the control parameters was predominantly linear. This array does allow for some nonlinearity, of a type called interaction between pairs of control parameters. However, if the relationship is highly nonlinear, the new settings found in step 3 might not be an improvement over the initial settings.

To guard against this possibility, a small follow-up experiment was conducted that included the new control parameter settings and, as a benchmark, the initial settings to confirm the prediction. The only differences in new and initial settings were those for nozzle position and rotation method. We conducted three independent test runs at each setting and calculated the mean and the log of the variance of epitaxial thickness from the results. Table 8.7 shows the average values of the mean, $\bar{y}$, and the log variance, $\log s^2$ at the new and the initial control parameter settings.

TABLE 8.7. CONFIRMATION EXPERIMENT

Experimental Run	Control Parameter								Average Value		
	A	B	C	D	E	F	G	H	$\bar{y}$	$\log s^2$	s^2
Initial settings	Osclt	678D4	1215	Low	57	1200	12	4	14.10	−0.845	0.143
New settings	Cont	678D4	1215	Low	57	1200	12	6	14.17	−1.244	0.057

The confirmation experiment's results show that the new settings reduced the variance of epitaxial thickness about 60 percent. There was almost no difference in the mean thickness at the two settings, because deposition time was not changed in the experiment.

In general, a confirmation experiment is essential before making a change to the manufacturing process, based on the results of a robust design experiment. Experiments that study many control parameters in a few runs are powerful tools for making improvements. However, many assumptions are made in planning these small experiments, and the confirmation experiment is insurance against incorrect assumptions.

8.5 FINDING MEAN-ADJUSTMENT PARAMETERS

In many manufacturing processes, one or more control parameters can be used to change a functional characteristic's mean, or average value, without affecting the process's variability. When these mean-adjustment control parameters (Leon, Shoemaker, and Kackar[3]) exist, we can choose settings of the other control parameters to minimize variability, while ignoring the mean. Then, we use the adjustment parameter to move the mean to target.

So, the key characteristics of a mean-adjustment parameter are a small effect on variability and a large effect on the mean.

In the epitaxial process, engineers use deposition time to control the mean epitaxial thickness. One goal of our robust design experiment was to reduce the process variability, but another was to verify that deposition time could indeed be used to adjust the mean epitaxial thickness without affecting its variability.

From Figure 8.3, we can see that deposition time has a smaller effect on the variance of epitaxial thickness than nozzle position and rotation method. To check that deposition time has a strong effect on the mean epitaxial thickness, we analyze $\bar{y}$ in the same way that log s^2 was analyzed in step 3.

Table 8.8 and Figure 8.4 show the average values of $\bar{y}$ at each control parameter setting. In Figure 8.4, notice that deposition time has the largest effect on the mean epitaxial thickness, $\bar{y}$.

TABLE 8.8. AVERAGE VALUE OF $\bar{y}$ AT DIFFERENT TEST SETTINGS

Control Parameter	Average $\bar{y}$		
	Test Setting 0	Test Setting 1	Difference
A. Susceptor-rotation method	14.4161	14.3616	−0.0545
B. Code of wafers	14.3610	14.4167	0.0556
C. Deposition temperature	14.4435	14.3342	−0.1094
D. Deposition time	14.8069	13.9709	−0.8359
E. Arsenic gas flow rate	14.4225	14.3552	−0.0674
F. Hydrochloric acid etch temperature	14.3589	14.4189	0.0600
G. Hydrochloric acid flow rate	14.4376	14.3401	−0.0975
H. Nozzle position	14.3180	14.4597	0.1417

After this analysis, we adopted the new control parameter settings for the epitaxial process and adjusted deposition time to make the mean epitaxial thickness equal 14.5 micrometers. Notice that, if we changed the target thickness, we could adjust deposition time accordingly, and the settings of the other control parameters would still give a small variance about this new target.

8.6 SUMMARY AND DISCUSSION

In this paper, we outlined the basic steps of the robust design method. Box, Hunter, and Hunter[1] provide more details on how to plan experiments, and Kackar[4] gives a thorough discussion of robust design. Also, the book by Taguchi and Wu[5] includes some examples of Japanese applications of this method.

Besides the examples described here and by Phadke,[6] the robust design method has been used to improve many other processes in AT&T. Examples include a window-photolithography process (Phadke and others[2]), the application of photoresist in hybrid IC manufacture, and reactive-ion etching and aluminum etching processes for 256K RAM manufacture.

Most AT&T applications of robust design have been to improve process designs rather than product designs. Although some details may be different, the general ideas apply to product design as well. For example, the Japanese used this method to improve the design of a truck steering mechanism Taguchi and Wu.[5]

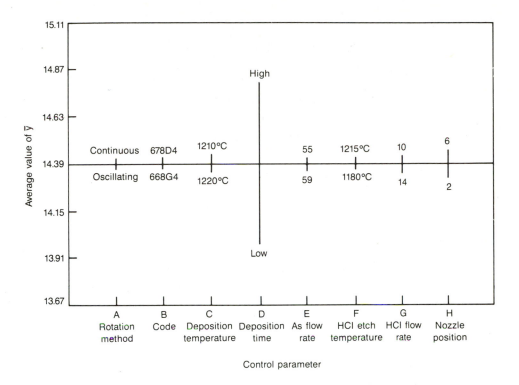

Figure 8.4. Average values of $\bar{y}$ at Each Setting of Control Parameters

However, for some products—such as electrical circuits—we know what function relates the product's functional characteristic to its control parameters and major sources of noise. For such products, other optimization methods that take advantage of the knowledge of this function and its derivatives may often be more efficient.

The robust design method uses an established statistical tool--the design of experiments--to help solve an important engineering problem: reducing variability caused by manufacturing variation. This use of designed experiments is Professor Genichi Taguchi's major contribution.

Statistically designed experiments have been used to improve industrial processes for more than 50 years (Box, Hunter, and Hunter[1]; Davies[7]; Daniel[8]), but most applications have focused on the mean values of the process's functional characteristics. In many process improvement applications, we can easily change the mean value using a mean-adjustment parameter. The yield of a manufacturing process is more closely linked to the process's variability, and robust design is a method for reducing that variability without increasing process cost.

ACKNOWLEDGMENTS

In this paper, we used two experiments conducted by AT&T's Technology Systems Group to illustrate the key ideas of robust design. We want to thank the following engineers, who were responsible for the experiments: D. G. Coult, S. M. Fisher, J. R. Mathews, D. H. Myers, and J. L. Pascuzzi.

REFERENCES

1. Box, G. E. P., W. G. Hunter, and J. S. Hunter. 1978. *Statistics for Experiments.* New York: John Wiley and Sons.

2. Phadke, M. S., R. N. Kackar, D. V. Speeney, and M. J. Grieco. 1983. "Off-line Quality Control in Integrated Circuit Fabrication Using Experimental Design." *The Bell System Technical Journal* vol. 62, no. 5 (May-June):1273-1309.
   (Quality Control, Robust Design, and the Taguchi Method; Article 6)

3. Leon, Ramon V., Anne C. Shoemaker, and Raghu N. Kackar. 1987. "Performance Measures Independent of Adjustment: An Explanation and Extension of Taguchi's Signal-to-Noise Ratios." *Technometrics* vol. 29, no. 3 (Aug.):253-265.
   (Quality Control, Robust Design, and the Taguchi Method; Article 12)

4. Kackar, R. N. 1985. "Off-line Quality Control, Parameter Design, and the Taguchi Method." *Journal of Quality Technology* **17** (Oct.):176-209.
   (Quality Control, Robust Design, and the Taguchi Method; Article 4)

5. Taguchi, G., and Y. Wu. 1979. *Introduction to Off-line Quality Control.* Japan: Central Japan Quality Control Association. Available from American Supplier Institute, 6 Parklane Boulevard, Suite 511, Dearborn, Mich. 48126.

6. Phadke, M. S. 1986. "Design Optimization Case Studies." *AT&T Technical Journal* vol. 65, no. 2 (Mar.-Apr.):51-68.
   (Quality Control, Robust Design, and the Taguchi Method; Article 10)

7. Davies, O. L. 1954. *The Design and Analysis of Industrial Experiments.* New York: Longman Inc.

8. Daniel, C. 1976. *Applications of Statistics to Industrial Experimentation.* New York: John Wiley and Sons.

9

TUNING COMPUTER SYSTEMS FOR MAXIMUM PERFORMANCE: A STATISTICAL APPROACH

William A. Nazaret and
William Klingler

9.1 INTRODUCTION

The work described in this paper has been motivated by two different, but related, problems that arise in the analysis of computer performance. The first is how to set an operating system's tunable parameters[1] to achieve the best response time for interactive tasks, given the computer's load conditions. The second is how to map the relationship among the different tunable parameters of the system and their impact on response time.

Although the second problem appears to be a generalization of the first, in practice the two appear in different contexts. The first problem is normally confronted by system administrators in their attempt to get the most performance out of the system on behalf of the users. In contrast, the second question is tackled mostly by system designers and performance analysts who are charged with modeling system performance under a variety of loads before the systems are actually handed to the customers. This activity is sometimes called "benchmarking."

The system administrator's goal is to optimize the response for important tasks in his/her organization under the particular load conditions imposed on the system by the users. Therefore we could say that this problem is "local" by nature.

---

1. Glossary equivalent is *control parameter*.

On the other hand, the responsibility of the system designer and performance analyst is to understand how the system reacts to changes in the tunable parameters for each of many different loads that are likely to be encountered on a system. In this sense the problem is rather "global".

A very important consequence of this distinction is that the measurements used for benchmarking are usually made under "simulated" loads that are designed to exercise the system in a manner totally determined by the experimenter. Tuning, however, uses data generated by the actual load of users on the system. This type of load is not under the complete control of the administrator or the experimenter. Despite the above differences, tuning and benchmarking have something in common—the necessity to experiment with different settings for the parameters in search of a configuration that yields the best results.

In this paper we present a systematic, cost effective approach to conducting these experiments. This approach makes use of statistical techniques to design experiments which yield, in many cases, information nearly equivalent to the one obtained by performing a complete exhaustive test. Our approach, although not new to statisticians, is now becoming popular among systems managers and performance analysts as an alternative to more traditional methods of experimentation.

Throughout the paper we use the UNIX[2] operating system as an example of a tunable operating system. However, the method is applicable to any operating system (or system in general) which allows the user the freedom to adjust its operating characteristics. In Section 9.2 we present an overview of UNIX tunable parameters and their potential impact on system response. Section 9.3 introduces the experimental problem by describing three experiments carried out on VAX[3] 780, 785 and 8600 machines respectively. Section 9.4 explains our statistical strategy to estimate the effect of the parameters on response time for certain tasks. In Section 9.5 we analyze the results of the experiments and show the improvement achieved after adjustment of the parameters according to these results. Finally, in Section 9.6 a critique of our approach is given along with some extensions.

9.2 WHAT DOES TUNING A COMPUTER MEAN?

Tuning a computer system is, in principle, not very different from tuning any system in general. It amounts to finding the setting of certain parameters to satisfy performance requirements. Hence, the performance requirements determine which settings are "optimal" for an application. For instance, a typical passenger car will

---

2. Registered trademark of AT&T.

3. Trademark of Digital Equipment Corporation.

be tuned to optimize fuel consumption and reduce emissions. In contrast, a racing car will be tuned to maximize speed at the expense of fuel economy.

The UNIX system allows fine tuning by giving the administrator the freedom to set the values of some kernel parameters at boot time. Additionally, one can exercise control over options that are not part of the kernel. Some of them relate to the hardware and some to software. Examples of these parameters and their significance for the run time environment are:

A. **System Buffers:** These are chunks of physical memory, typically 1024 kilobytes in size, which are used by the operating system to keep recently used data in hope that it might be used shortly afterwards. Increasing the number of these buffers improves the "hit ratio" on this cache up to a point. On the other hand, an excessive amount of these buffers can hurt performance since it takes away memory space from the users.

B. **Sticky Processes:** There is a bit associated with the permissions on an executable file that will cause its text segment to be stored in contiguous blocks on the swapping device. Commands which are frequently invoked (specially those with large images) ought to have the sticky bit set so that every time they are invoked their code can be brought into memory as easily as possible by the system. Systems in which this is not done usually suffer from chronic disk I/O bottleneck and the resulting degradation in response time. The number and kind of commands with the sticky bit set is a tunable parameter.

C. **Paging Daemon Parameters:** In virtual memory implementations of UNIX, memory used by processes is assigned on a per-page basis. A page is just a piece of the code usually 512 or 1024 kilobytes long. The paging daemon is a system process whose responsibility is to free up memory by reclaiming space occupied by pages which are no longer in use. A process can also be stripped of its pages if its total CPU time exceeds a given value. How often the daemon runs, how many simultaneous active pages a process can have and the maximum CPU time quota before a process is swapped out are tunable parameters.

D. **File System Organization:** This is a highly installation dependent parameter. The idea is to distribute the system and user files among the available disks in a way that the load on each of them is approximately the same. When one of the disks is overloaded I/O waits increase and response is degraded.

E. **CPU Assist Devices:** Some types of hardware allow the possibility of adding coprocessors or add-on boxes to relieve the CPU of mundane chores. Some examples are terminal I/O assist devices and troff coprocessors. The use and number of such devices can be subjected to tuning.

F. **Main Memory:** By allowing minor changes in the amount of physical memory on the system it is possible to detect whether increasing memory size will help to enhance the performance of the system. This is helpful to know before committing any resources into buying the additional boards.

9.3 THREE CASE STUDIES

To illustrate the methodology we will describe three experiments carried out on VAX 780, 785 and 8600 systems respectively, running under the UNIX System V Operating System at the Quality Assurance Center of AT&T Bell Laboratories.

The first of the three was conducted with primarily a tuning orientation, and is somewhat similar to the one reported in Pao, Phadke, and Sherrerd.[1] Our goals were to improve response on a system whose performance was becoming unbearable and to overcome some questionable aspects of the experiment in Pao, Phadke, and Sherrerd.[1] Among these aspects we targeted:

- Duration of the experiments: The experiment described in Pao, Phadke, and Sherrerd[1] lasted more than three months. We believe that any approach to tuning which takes this long to produce results has very limited practical value. Therefore, one of our goals was to find ways to obtain useful results in a reasonable amount of time.

- Stationarity of the load: A second goal was to ensure that the load conditions were reasonably stationary. We wanted to avoid a situation in which it was not possible to determine whether improvements in performance were due to the tuning or to a decreased load level on the system (this is essentially what happened in Pao, Phadke, and Sherrerd[1]).

The tunable parameters considered in this experiment were: file system organization, main memory size, system buffers, sticky processes, and two type of CPU assist devices-KMC's (terminal I/O processors) and PDQ's (troff co-processors).

The second experiment on a VAX 785 system included only 5 of the 6 parameters from the VAX 780 experiment. Sticky processes were dropped out of consideration because we already had considerable prior knowledge about how to handle them. The experiment was designed to allow us to estimate, in addition to the main effects of the factors,[4] some of the interactions among them. Therefore it resulted in a larger number of trials. Our intentions here went beyond tuning the system. We also wanted to assess the merits of fancier (more expensive) design plans relative to simple plans like the one used in Experiment One. We will elaborate on this in the next section when we discuss the experimental strategy.

---

4. Glossary equivalent is *control parameter*.

The third experiment differed from the previous ones in a very important characteristic. The load used was not a "live" load but rather a simulated load designed to be representative of the type of load we expected this system to be subjected to. The factors considered were the same as in the second experiment except that KMC's were not included (they are not necessary in this new VAX 8600 model). By using a simulated load, the experimental conditions were completely under our control and thus we expected to get more statistically reliable results. However, there was no previous experience in using our method in this context and little was known about the validity of the conclusions under actual load conditions. What we learned from this is discussed in Section 9.6.

9.4 THE EXPERIMENTAL STRATEGY

Before basic planning for the experiment can be done, we need to choose the various factor levels to be tried. Since the set of factors considered in experiment one (VAX-780) contains the ones for the other two experiments it will suffice to describe that case. Figure 9.1 shows the levels for each of the six factors.

		Levels	
Factors	**1**	**2**	**3**
File System Distribution	A	B	C
KMC's	NO	YES	–
Memory Size	6 Mb	7 Mb	8 Mb
System Buffers	Low	Medium	High
Sticky Processes	10	20	30
PDQ's	None	One	Two

Figure 9.1. VAX-780

In Figure 9.1 the amount of system buffers space allocated depends on the total size of memory. Hence, Low, Medium and High represent a different fraction of memory for each of the three memory sizes. We determined the amount of memory assigned to these buffers using the formula

$$Sysbuff = C_L + .2 \cdot K_L,$$

where C_L is 1.0, 1.2 and 1.4 Megabytes when L is Low, Medium, and High respectively. Similarly, K_L is 0,1 or 2.

The choice of levels for the factors is highly installation dependent and must be done taking into account both the characteristics of the load and prior knowledge about how changes in these parameters are supposed to affect the system's response (AT&T[2] is an excellent reference for this). A common strategy starts from the current settings and introduces some variations around them. The size of this variation ranges from modest to large. Minor variations defeat the purpose of the experiment. One exception to this strategy occurs when the current setting of one of the parameters is clearly wrong (non-optimal). In Figure 9.1 we have such an example. In experiment 1 the number of sticky processes before the experiment was 40. However, these processes had been chosen without regard to their size or frequency of invocation. Instead, we chose the 10, 20 and 30 most popular commands as the levels for this experiment.

Next we had to decide on the measures to assess the performance of the system under the different experimental conditions. A common choice is the time the system needs to execute a script containing tasks important to the organization. For instance, in a text processing organization such a script would consist of formatting a document. The four measures we typically use are: trivial time, edit time, troff time and c-compile time. Among these, only "trivial time" is not self-explanatory. This is just the response time for a command that involves no interaction between the user and the system (e.g., the "date" command). It gives a measure of instantaneous response time.

Deciding on a sampling plan for the experiment is a crucial and difficult task. As we noted above, we wanted to reduce the total time for the experiment, as much as possible, without compromising the integrity of the methodology. After careful study of the load in the VAX-780 and VAX-785 systems we decided that it was safe to use a day as the basic duration of a run. A day is a natural unit because it allows you to setup the system from one run to the next during off-hours, sparing the users any inconvenience. This choice is also minimal in the sense that the runs lasting less than a day would interfere with the normal functioning of the system. More importantly, it would make comparisons among runs invalid due to the within day variations of the load (peak hours). For some installations even the one day test unit could be a problem if the load varies significantly by day of the week.

In the third of the experiments (VAX-8600) a whole run took only about an hour as opposed to a whole day. This, of course, was due to the use of a simulated load.

Response times were measured at evenly spaced intervals during the run. It is important not to oversample since the load caused by the timing programs and the timed scripts could interfere with the users.

The most important aspect of the whole experimental strategy is the choice of level combinations to be tried during the experiment. A complete exhaustive search would most likely give us the right answer. However, it is obvious that the time to do this would be prohibitive. For the VAX-780 example above (Figure 9.1)

180

it would take about 729 days to run an all-combinations experiment! Even in the case of the third experiment (with simulated loads) the administrative overhead is overwhelming. It requires re-booting the system 81 times! Rather, we selected an array of factor combinations that allows us to test simultaneously all the factors in a limited number of trials. Such arrays are documented extensively in the statistical literature (see Taguchi and Wu[3] and Box, Hunter, and Hunter[4]). For instance, the design used in Experiment 1 (VAX-780) is shown in Figure 9.2. It was constructed using an orthogonal array known as the L_{18}, consisting of only 18 runs. Figure 9.2 shows the factor settings for each of the 18 runs.

Run #	File Sys	KMC's	Memory Size	Buffers Space	Sticky Process	PDQ's
1	A	No	7 Mb	1.4	10	None
2	A	No	6 Mb	1.0	20	One
3	A	No	8 Mb	1.8	30	Two
4	A	Yes	7 Mb	1.2	30	One
5	A	Yes	6 Mb	1.4	10	Two
6	A	Yes	8 Mb	1.6	20	None
7	B	No	7 Mb	1.4	20	One
8	B	No	6 Mb	1.0	30	Two
9	B	No	8 Mb	1.8	10	None
10	B	Yes	7 Mb	1.6	30	None
11	B	Yes	6 Mb	1.2	10	One
12	B	Yes	8 Mb	1.4	20	Two
13	C	No	7 Mb	1.6	10	Two
14	C	No	6 Mb	1.2	20	None
15	C	No	8 Mb	1.4	30	One
16	C	Yes	7 Mb	1.6	20	Two
17	C	Yes	6 Mb	1.0	30	None
18	C	Yes	8 Mb	1.6	10	One

Figure 9.2. VAX-780

For the second experiment we used another orthogonal array known as the L_{27} (see Taguchi and Wu[3]), consisting of 27 runs. The increased size of the experiment, as we mentioned previously, was deliberately planned to allow us to estimate together with the main effects, the interactions between memory size and the other four factors. Finally, the controlled experiment (VAX 8600) with only four factors was run following a plan based on a fraction of a 3^4 array consisting of just 9 runs.

The advantages of using design plans like the ones above are:

- They provide an average picture over the whole parameter space.

- The estimate of the effect of any of the factors is orthogonal with respect to those of the other factors. (i.e. the main effect of any individual factor can be interpreted without regard for those of the other factors).

- Under certain conditions they yield information approximately equivalent to what you would obtain by using a much larger experiment.

There is of course a price to pay for this. These plans achieve the reduction in size of the experiment by deliberately confounding the main effect of the factors with the "joint" effect or interaction of some of the other factors. Therefore, they rely on the size of the main effects to be dominant.

In spite of the above we advocate a strategy based on choosing a highly frac-tioned array because, at the very least, it provides a most inexpensive starting point from which we can always obtain very useful information about the parame-ters. In particular, these experiments can be extended, if the data suggest that higher order effects (interactions) might be important. In practice, we have seldom had to exceed one iteration in this cycle. Usually the improvement achieved by implementing the predicted optimal setting is sufficient to make further experi-ments an unnecessary effort.

9.5 DATA ANALYSIS

Due to space constraints we cannot present summaries and analyses of the data for each of the three experiments. We can however show a selected subset of plots summarizing our findings on the relationship between factors and perfor-mance. We will also show plots illustrating the outcomes of re-setting the parame-ters to levels suggested by the experimental data.

We analyzed the data first using mean square response and then mean response as performance measures. The former has the advantage of incorporating the average response as well as its variability, while the latter is easier to interpret. Since the conclusions obtained using either measure were equivalent for these experiments, we show the results based on mean response time. Figure 9.3 shows (in seconds) the estimated effects that each of the four parameters in Experiment number 3 (VAX 8600) had on mean c-compile time.

Recommendations we made based on these data are:

- The machine should be run at its current level of 8 Megabytes. The gradient information in Figure 9.3 does not suggest significant additional gains if another 4 Megabytes are added (memory can be bought in 4 Mb units).

- The number of system buffers can be set to low, which means about 1.4 Megabytes of memory for the system and the rest for the users.

- File system organization C is advantageous over either A or B.
- No PDQ's should be used.

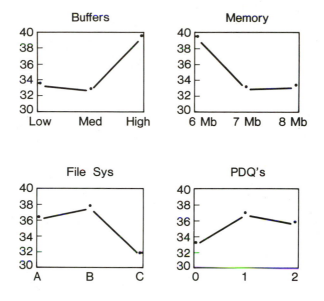

Figure 9.3. C-compile

These conclusions are valid for this particular response measure and for this system. There is no reason why they would be the same for another response measure (like edit time) even under the same load. For instance, one could argue that not adding PDQ's to the system could hurt text processing performance and this could very well be the case. In general the answer depends on the relative level of capability of the processor to handle the load. As a matter of fact, to our surprise we have seen cases in which adding PDQ's even hurts "troff" response time. An explanation for this puzzling event is that the VAX processor is several times more powerful than the microprocessor which drives the PDQ's. Therefore every time a text processing job is sent to the PDQ when the CPU could indeed have handled it, a loss in performance results. The results of the first experiment recommended the inclusion of one or more PDQ's in the system configuration. For this particular system the load was so heavy that there was little change in troff performance by adding or excluding the PDQ's, while there was a positive effect on edit and trivial response times upon adding them.

A way to check the gains in performance after tuning is to run back to back confirmatory runs under the old and new settings. The results for one of the three systems is given in Figure 9.4. In the graphs the dotted curve represents the response time before the experiment and the solid curve represents the response after the tuning. We see that both trivial and response times were reduced considerably after tuning the system.

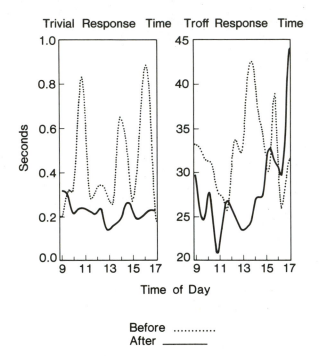

Figure 9.4. Comparison of Response Times

9.6 CONCLUSIONS

We were able to reduce response time for several typical tasks in all three systems. For the first system, we reduced response by 38% (on average), an important gain in a system that was considered hopeless. In the second system, we also discover that blind use of PDQ's could lead to loss in performance for text processing jobs. The evaluation of the results for the third system is still underway. Early data seem to indicate that the configuration recommended by the simulated load experiment enables the machine to handle the real load rather easily.

We also have a much better assessment of the real usefulness of these experiments for both tuning and benchmarking. The results of back to back confirmatory runs (a week each) showing substantial reductions in response time for both of the tuning experiments, indicate that it is possible to use this approach successfully for periodical system tuning. We can not however claim that it will succeed in general. Instead we can say that as long as attention is confined to only a few factors (therefore keeping the duration short) and the load is relatively stable, the method will help to run your system better. The experience with simulated loads convinced us that the usefulness of this approach in benchmarking studies is even greater. In fact, we are currently using our approach to find the functional relation between the parameters of the UNIX System V Virtual Memory Management scheme and response time.

Finally, with regard to the problem of ignoring the interaction effects among some of the factors we would like to point out that, in theory, there is a danger. However, in practical situations where time and resources are limiting factors, the potential benefits of using statistically planned experiments (even if they do not allow for the estimation of interactions), makes our strategy a worthy one.

REFERENCES

1. Pao, T. W., M. S. Phadke, and C. S. Sherrerd. 1985. "Computer Response Time Optimization Using Orthogonal Array Experiments." IEEE International Communication Conference, Chicago, *Conference Record* vol. 2:890-895.

2. AT&T Information Systems. n. d. *UNIX System V Tuning and Configuration Guide.*

3. Taguchi, G., and Y. Wu. 1979. *Off-line Quality Control.* Tokyo: Central Japan Quality Control Association.

4. Box, G. E. P., W. Hunter, and J. S. Hunter, 1978. *Statistics for Experimenters.* New York: John Wiley and Sons.

10

DESIGN OPTIMIZATION CASE STUDIES

Madhav S. Phadke

10.1 AN ECONOMIC AND TECHNOLOGICAL CHALLENGE

A goal of product or process design is to provide good products under normal manufacturing conditions and under all working conditions throughout intended life. Further, the cost of making the product (including development and manufacturing) must be low, and the development must be speedy to meet the market needs. Achieving these goals is an economic and technological challenge to the engineer. A systematic and efficient way to meet this challenge is the method of design optimization for performance, quality, and cost, developed by Genichi Taguchi of Japan. Called "robust design," the method has been found effective in many areas of engineering design in AT&T, Ford, Xerox, ITT, and other American companies. In this paper we will describe the basic concepts of robust design and describe the following two applications in detail:

- A router bit life improvement study
- Optimization of a differential operational amplifier circuit.

The first application illustrates how, with a very small number of experiments, highly valuable information can be learned about a large number of variables for improving the life of router bits used for cutting printed wiring boards from panels. The study also illustrates how product life improvement projects should be organized for efficiency. This case study involved conducting hardware experiments, whereas in the second case study a computer simulation model was used. The second application shows the optimization of a differential op-amp circuit to minimize the dc offset voltage. This is accomplished primarily by moving

Reprinted with permission from the *AT&T Technical Journal*. Copyright 1986 AT&T.

the center point of the design, which does not add to the cost of making the circuit. Reducing the tolerance could have achieved similar improvement, but at a higher manufacturing cost!

10.2 PRINCIPLES OF ROBUST DESIGN

Genichi Taguchi views the quality of a product in terms of the total loss incurred by a society from the time the product is shipped to the customer. The loss may result from undesirable side effects arising from the use of the product and from the deviation of the product's function from the target function. What is novel about this view is that it explicitly includes the cost to the customers and the notion that even products that meet the "specification limits" can impart loss due to nonoptimum performance. For example, the amplification level of a public telephone set may differ from cold winter to hot summer; it may differ from one set to another; also it may deteriorate over a period of time. A consequence of this variation is that a user of the phone may not hear the conversation well and that an expensive compensation circuit may have to be provided. The quadratic loss function can estimate with reasonable accuracy the loss due to functional variation in most cases. For a broad description of the principles of robust design see Taguchi,[1] Taguchi and Phadke,[2] Phadke,[3] and Kackar.[4]

10.2.1 SOURCES OF VARIATION

Robust design is aimed at reducing the loss due to variation of performance from the target. In general, a product's performance is influenced by factors that are called noise factors.[1] There are three types of noise factors:

1. *External*—factors outside the product, such as load conditions, temperature, humidity, dust, supply voltage, vibrations from nearby machinery, human errors in operating the product, and so forth.

2. *Manufacturing imperfection*—the variation in the product parameters from unit to unit, inevitable in a manufacturing process. For example, the value of a particular resistor in a unit may be specified as 100 kilohms, but in a particular unit it turns out to be 101 kilohms.

3. *Deterioration*—when the product is sold, all its performance characteristics may be right on target, but as years pass by, the values or characteristics of individual components may change, leading to product performance deterioration.

---

1. Glossary equivalent is *noise*.

One approach to reducing a product's functional variation is to control the noise factors. For the telephone set example, it would mean to reduce allowable temperature range, to demand tighter manufacturing tolerance or to specify low-drift parameters. These are costly ways to reduce the public telephone set amplification variation. What then is a less costly way? It is to center the design parameters in such a way as to minimize sensitivity to all noise factors. This involves exploiting the nonlinearity of the relationship between the control factors,[2] the noise factors, and the response variables. Here, control factor means a factor or parameter over which the designer has direct control and whose level or value is specified by the designer.

Note that during product design one can make the product robust against all three types of noise factors described above, whereas during manufacturing process design and actual manufacturing one can reduce variation due to manufacturing imperfection, but can have only minor impact on variation due to the other noise factors. Once a product is in the customer's hand, warranty service is the only way to address quality problems. Thus, a major portion of the responsibility for the quality and cost of a product lies with the product designers and not with the manufacturing organization.

10.2.2 STEPS IN PRODUCT/PROCESS DESIGN

Product and process design are complex activities involving many steps. Three major steps in designing a product or a manufacturing process are system design, parameter design or design optimization, and tolerance design. System design consists of arriving at a workable circuit diagram or manufacturing process layout. The role of parameter design or design optimization is to specify the levels of control factors that minimize sensitivity to all noise factors. During this step, tolerances are assumed to be wide so that manufacturing cost is low. If parameter design fails to produce adequately low functional variation of the product, then during tolerance design, tolerances are selectively reduced on the basis of cost effectiveness.

10.2.3 THE DESIGN OPTIMIZATION PROBLEM

A product or a process can be represented by a block diagram (see Figure 10.1) proposed by Taguchi and Phadke.[2] The diagram can also be used to represent a manufacturing process or even a business system. The response is

---

2. Glossary equivalent is *control parameter*.

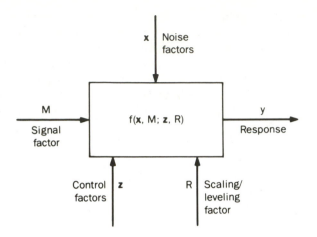

Figure 10.1. Block Diagram of a Product or Process

represented by y. The factors that influence the response can be classified into four groups as follows:

1. Signal factors (M): These are the factors that are set by the user/operator to attain the target performance or to express the intended output. For example, the steering angle is a signal factor[3] for the steering mechanism of an automobile. The speed control setting on a fan and the bits 0 and 1 transmitted in communication systems are also examples of signal factors. The signal factors are selected by the engineer on the basis of engineering knowledge. Sometimes two or more signal factors are used in combination; for example, one signal factor may be used for coarse tuning and one for fine tuning. In some situations, signal factors take on a fixed value, as in the two applications described in this paper.

2. Control factors (z): These are the product design parameters whose values are the responsibility of the designer. Each of the control factors can take more than one value; these multiple values will be referred to as levels or settings. It is the objective of the design activity to determine the best levels of these factors. A number of criteria may be used in defining the best levels; for example, we would want to maximize the stability and robustness of the design while keeping the cost to a minimum. Robustness is the insensitivity to noise factors.

---

3. Glossary equivalent is *signal parameter*.

3. Scaling/leveling factors[4] (R): These are special cases of control factors that can be easily adjusted to achieve a desired functional relationship between the signal factor and the response y. For example, the gearing ratio in the steering mechanism can be easily adjusted during the product design phase to achieve the desired sensitivity of the turning radius to a change in the steering angle. The threshold voltage in digital communication can be easily adjusted to alter the relative errors of transmitting 0's and 1's. Scaling/leveling factors are also known as adjustment design parameters.

4. Noise factors (x): Noise factors, described earlier, are the uncontrollable factors and the factors that we do not wish to control. They influence the output y and their levels change from one unit of the product to another, from one environment to another, and from time to time. Only the statistical characteristics of the noise can be known or specified, not their actual values.

Let the dependence of the response y on the signal, control, scaling/leveling, and noise factors be denoted by

$$y = f(\mathbf{x}, M; \mathbf{z}, R).$$

Conceptually, the function f consists of two parts: $g(M; \mathbf{z}, R)$, which is the predictable and desirable functional relationship between y and M, and $e(\mathbf{x}, M; \mathbf{z}, R)$, which is the unpredictable and less desirable part. Thus,

$$y = g(M; \mathbf{z}, R) + e(\mathbf{x}, M; \mathbf{z}, R).$$

In the case where we desire a linear relationship between y and M, g must be a linear function of M. All nonlinear terms will be included in e. Also, the effect of all noise variables is contained in e.

The design optimization can usually be carried out in two steps:

- Find the settings of control factors to maximize the predictable part while simultaneously minimizing the unpredictable part. This can be accomplished through an optimization criterion called signal-to-noise (S/N) ratio. In this step the variability of the functional characteristic is minimized.

- Bring the predictable part, $g(M; \mathbf{z}, R)$, on target by adjusting the scaling/leveling factors.

---

4. Glossary equivalent is *adjustment parameter*.

Design problems come in a large variety. For a classification of design problems and the selection of S/N ratios see Taguchi and Phadke.[2]

10.3 ROUTER BIT LIFE IMPROVEMENT

10.3.1 THE ROUTING PROCESS

Typically, AT&T printed wiring boards are made in panels of 18×24 inch size. Appropriately sized boards, say 8×4 inches, are cut from the panels by stamping or by routing. A benefit of routing is that it gives good dimensional control and smooth edges, thus reducing friction and abrasion during circuit pack insertion. However, when the router bit gets dull, it produces excessive dust, which cakes on the edges and makes them rough. In such cases, a costly cleaning operation is necessary to smooth the edges. But changing the router bits frequently is also expensive.

The routing machine has four spindles, all synchronized in rotational speed, horizontal feed ($x - y$ feed) and vertical feed (in-feed). Each spindle does the routing operation on a separate stack of panels. Two to four panels are usually stacked for cutting by a spindle. The cutting process consists of lowering the spindle to an edge of a board, cutting the board all around using the $x - y$ feed of the spindle, and then lifting the spindle. This is repeated for each board on the panel.

Our objective in this experiment was to increase the life of the router bits, primarily in regard to the onset of excessive dust formation. The dimensions of the board were well in control and were not an issue.

10.3.2 SELECTION OF CONTROL FACTORS AND THEIR LEVELS

Selecting appropriate control factors and their alternate settings is an important aspect of optimization. Prior knowledge and experience about the process is used in this selection. The alternate settings are called levels. It is a good practice to choose these levels wide apart so that a broad design space is studied in one set of experiments and there is a potential for major improvement. For the routing process, the eight control factors listed in Table 10.1 were chosen.

Suction is used around the router bit to remove the dust as it is generated. Obviously, higher suction could reduce the amount of dust retained on the boards. The starting suction was 2 inches of mercury—the maximum available for the pump. We chose 1 inch of mercury as the alternate level, with the plan that if a significant difference in the dust was noticed, we would invest in a more powerful pump. Related to the suction are suction foot and the depth of backup slot. The suction foot determines how the suction is localized near the cutting point. Two types of suction foot were chosen: solid ring and bristle brush. Underneath the

TABLE 10.1. CONTROL FACTORS FOR THE ROUTING PROCESS

		Level			
	Factor	1	2	3	4
A	Suction (in of Hg)	1	2*		
B	$x - y$ feed (in/min)	60*	80		
C	In-feed (in/min)	10*	50		
D	Type of bit	1	2	3	4*
E	Spindle position†	1	2	3	4*
F	Suction foot	SR	BB*		
G	Stacking height (in)	3/16	1/4*		
H	Depth of slot (mils)	60*	100		
I	Speed (rpm)	30,000	40,000*		

* Denotes starting condition for the factors.

† Spindle position is not a control factor. In the interest of productivity, all four spindle positions must be used.

panels being routed is a backup board. Slots are precut in the backup board to provide air passage and a place for dust to temporarily accumulate. The depth of the slots was a control factor in this study.

Stack height and $x - y$ feed are control factors related to the productivity of the process; that is, they determine how many boards are cut per hour. The 3/16-inch stack height means three panels were stacked together, while 1/4-inch stack height means four panels were stacked together. The in-feed determines the impact force during the lowering of the spindle for starting to cut a new board. It could influence the life of the bit by causing breakage or damage to the point. Four different types of router bits made by different manufacturers were used. The router bits varied in cutting geometry in terms of the helix angle, the number of flutes, and the point. Spindle position was not a control factor. All spindle positions must be used in production, otherwise productivity would suffer. It was included in the study so that we could find best settings of the control factors to work well with all four spindles.

In addition to the variation from spindle to spindle, the noise factors for the routing process are the bit-to-bit variation, the variation in material properties within a panel and from panel to panel, the variation in the speed of the drive motor, and similar factors.

10.3.3 THE ORTHOGONAL ARRAY EXPERIMENT

The full factorial experiment to explore all possible factor-level combinations would require $4^2 \times 2^7 = 2048$ experiments. Considering the cost of material, time, and availability of facilities, the full factorial experiment is prohibitively large. However, it is unnecessary to perform the full factorial experiment because processes can usually be characterized by relatively few parameters. An orthogonal array design with 32 experiments was created from the L_{16} array and the linear graphs given in Taguchi and Wu.[5] The array appears in Table 10.2. This design allowed us to obtain uncorrelated estimates of the main effect of each control factor as well as of the spindle position, and the interactions between $x - y$ feed and speed, in-feed and speed, stack height and speed, and $x - y$ feed and stack height.

TABLE 10.2. EXPERIMENT DESIGN AND OBSERVED LIFE FOR THE ROUTING PROCESS

Experi- ment No.	Suction A	$x - y$ feed B	In-feed C	Bit D	Spindle E	Suction foot F	Stack height G	Depth H	Speed I	Observed life*
1	1	1	1	1	1	1	1	1	1	3.5
2	1	1	1	2	2	2	2	1	1	0.5
3	1	1	1	3	4	1	2	2	1	0.5
4	1	1	1	4	3	2	1	2	1	17.5
5	1	2	2	3	1	2	2	1	1	0.5
6	1	2	2	4	2	1	1	1	1	2.5
7	1	2	2	1	4	2	1	2	1	0.5
8	1	2	2	2	3	1	2	2	1	0.5
9	2	1	2	4	1	1	2	2	1	17.5
10	2	1	2	3	2	2	1	2	1	2.5
11	2	1	2	2	4	1	1	1	1	0.5
12	2	1	2	1	3	2	2	1	1	3.5
13	2	2	1	2	1	2	1	2	1	0.5
14	2	2	1	1	2	1	2	2	1	2.5
15	2	2	1	4	4	2	2	1	1	0.5
16	2	2	1	3	3	1	1	1	1	3.5
17	1	1	1	1	1	1	1	1	2	17.5
18	1	1	1	2	2	2	2	1	2	0.5
19	1	1	1	3	4	1	2	2	2	0.5
20	1	1	1	4	3	2	1	2	2	17.5
21	1	2	2	3	1	2	2	1	2	0.5
22	1	2	2	4	2	1	1	1	2	17.5
23	1	2	2	1	4	2	1	2	2	14.5
24	1	2	2	2	3	1	2	2	2	0.5
25	2	1	2	4	1	1	2	2	2	17.5
26	2	1	2	3	2	2	1	2	2	3.5
27	2	1	2	2	4	1	1	1	2	17.5
28	2	1	2	1	3	2	2	1	2	3.5
29	2	2	1	2	1	2	1	2	2	0.5
30	2	2	1	1	2	1	2	2	2	3.5
31	2	2	1	4	4	2	2	1	2	0.5
32	2	2	1	3	3	1	1	1	2	17.5

* Life was measured in hundreds of inches of movement in $x - y$ plane. Tests were terminated at 1700 inches.

This information about the effect of control factors was used to decide the best setting for each factor. The 32 experiments were arranged in groups of four so that for each group there was a common speed, $x - y$ feed, and in-feed, and the four experiments in each group corresponded to four different spindles. Thus each group constituted a machine run using all four spindles, and the entire experiment could be completed in eight runs of the routing machine.

The study was conducted with one bit per experiment; thus a total of only 32 bits was used. During each machine run, the machine was stopped after every 100 inches of cut (100 inches of router bit movement in the $x - y$ plane) so that the amount of dust could be inspected. If the dust was beyond a predetermined level, the bit was recorded as failed. Also, if a bit broke, it was obviously considered to have failed. Otherwise, it was considered as having survived.

Before the experiment was started, the average bit life was estimated at around 850 inches. Therefore, to save time, each experiment was stopped at 1700 inches of cut, which is twice the estimated original average life, and the survival or failure of each bit was recorded.

Table 10.2 gives the experimental data in hundreds of inches. A reading of 3.5 means that the bit failed between 300 and 400 inches. Other readings have similar interpretations, except the reading of 17.5, which means survival beyond 1700 inches, the point where the test was terminated. There are 14 readings of 0.5, indicating extremely unfavorable conditions. There are eight cases of life equal to 17.5, indicating very favorable conditions. During experimentation, it is important to take a broad range for each control factor so that roughly equal numbers of favorable and unfavorable conditions are created. In this way, much can be learned about the optimum settings of control factors.

10.3.4 ANALYSIS OF THE LIFE DATA AND RESULTS

Two simple and separate analyses of the life data were performed to determine the best level for each control factor. The first analysis was to determine the effect of each control factor on the failure time. The second analysis was performed to determine how changing the level of each factor changes the survival probability curve (life curve).

The first analysis was performed by the standard procedure for fractional factorial experiments given in Hicks[6] and Box, Hunter and Hunter.[7] In this analysis of variance, the effect of censoring was ignored. The results are plotted in Figures 10.2 and 10.3. The following conclusions are apparent from the plots:

- 1-inch suction is as good as 2-inch suction.

- Slower $x - y$ feed gives longer life.

- The effect of in-feed is small.

- The starting bit is the best of the four bit types.

- The differences among the spindle positions are small.

- A solid ring suction foot is better than the bristle brush type.

- Lowering the stack height makes a large improvement. This change, however, raises machine productivity issues.

- The depth of slot in the back-up material has negligible effect.

- Higher rotational speed gives improved life. If the machine stability permits, even higher speeds should be tried in the next cycle of experiments.

The only two-factor interaction that is large is the stack height versus speed interaction. However, the optimum settings of these factors suggested by the main effects are consistent with those suggested by the interaction. The best factor level combination suggested by the above results and the starting factor level combination are tabulated in Figures 10.2 and 10.3.

Using the linear model of Taguchi and Wu[5] and taking into consideration only the terms for which the variance ratio is large, that is the factors B, D, F, G, I and interaction $I \times G$, we can predict the router bit life under starting, optimum, or any other combination of factor settings. The predicted life under the starting conditions is 860 inches and under optimum conditions is 2200 inches. Because of the censoring at 1700 inches, these predictions are obviously likely to be on the low side. The prediction for optimum conditions especially is likely to be much less than the realized value. From the machine logs, the router bit life under starting conditions was found to be 900 inches, while the confirmatory experiment under optimum conditions yielded an average life in excess of 4150 inches.

In selecting the best operating conditions for the routing process, one must consider the overall cost, which includes not only the cost of router bits but also the cost of machine productivity, the cost of cleaning the boards if needed, and so forth. Under the optimum conditions listed in Figure 10.3, the stack height is 3/16 inch as opposed to 1/4 inch under the starting conditions. This means three panels are cut simultaneously instead of four panels. The lost machine productivity due to this change can however be made up by increasing the $x - y$ feed. If the $x - y$ feed is increased to 80 in/min, the productivity of the machine would get back approximately to the starting level. The predicted router bit life under these alternative optimum conditions is 1700 inches, which is twice the predicted life for starting conditions. Thus a 50 percent reduction in router bit cost can be achieved while still maintaining machine productivity. An auxiliary experiment would be needed to precisely estimate the effect of $x - y$ feed under the new settings of all other factors. This would enable us to make an accurate economic analysis.

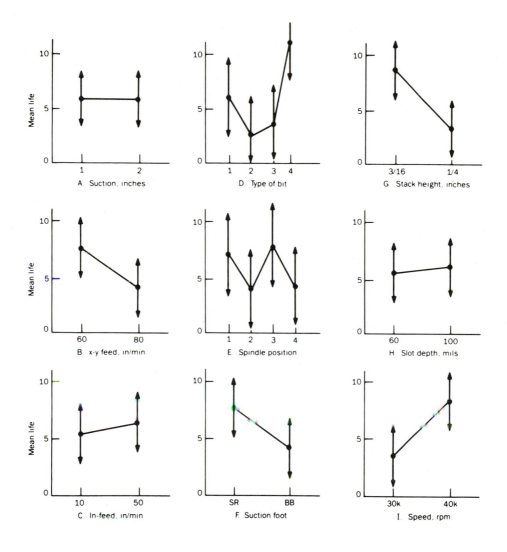

Figure 10.2. Average Factorial Effects. Mean life is given in hundreds of inches. The 2 σ limits are also shown.

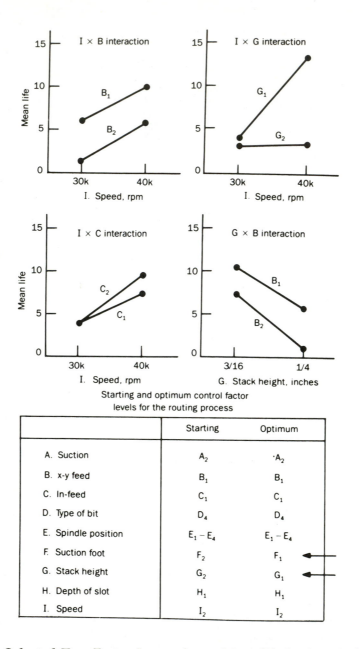

Figure 10.3. **Selected Two-Factor Interactions**. Mean life is given in hundreds of inches. The 2 σ limits are also shown. The table indicates the starting and optimum control factor levels for the routing process.

10.3.5 SURVIVAL PROBABILITY CURVE

The life data can also be analyzed in a different way, by the minute analysis method described in Taguchi and Wu,[5] to construct the survival probability curves for the levels of each factor. To do so, we look at every 100 inches of cut and note which router bits have failed and which have survived. Treating this as 0-1 data, we can determine factorial effects by the standard analysis method. Thus for suction levels A_1 and A_2, the survival probabilities at 100 inches of cut were estimated to be 0.44 and 0.69. Likewise the probabilities are estimated for each factor and also for each time period: 100 inches, 200 inches, etc. These data, plotted in Figure 10.4, graphically display the effects of factor level changes on the entire life curve. The conclusions from these plots are consistent with the conclusions from the analysis described earlier.

Plots like Figure 10.4 can be used to determine the entire survival probability curve under a new set of factor level combinations such as the optimum combination. See Taguchi and Wu[5] the method of calculation.

Notice that in this method of determining life curves, no assumption was made regarding the shape of the curve—whether it follows a Weibull or a lognormal distribution, for example. Also, the total amount of data needed to come up with the life curves is small. In this example it took only 32 samples to determine the effects of eight control factors. For a single good fit of a Weibull distribution one typically needs several tens of observations. So the approach used here can be very beneficial for reliability improvement projects.

There are, of course, some caveats. First, as in any fractional factorial experiment, one needs to guard against the interactions among the various control factors. But this difficulty can be overcome through the confirmatory experiment. Second, the method for determining the statistical significance of the differences between the life curves for different factor levels needs more research.

10.4 CIRCUIT DESIGN OPTIMIZATION

The differential operational amplifier circuit is commonly used in telecommunications. An example is as a preamplifier in coin telephones, where it is expected to function over a wide temperature range. An important characteristic of this circuit is its offset voltage. (If the offset voltage is large, then the circuit cannot be used over long loops between the central office and the telephone). So the optimization objective was to minimize the offset voltage. The balancing property of the circuit makes the offset voltage small under nominal conditions. What needs to be minimized is the effect of tolerances and temperature variation on the offset voltage.

In the circuit diagram (see Figure 10.5), there are two current sources, five transistors, and eight resistors. This differential operational amplifier circuit is made as part of a larger integrated circuit.

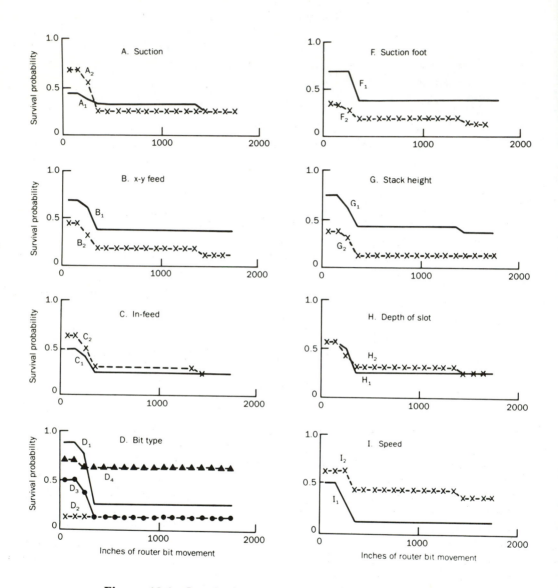

Figure 10.4. Survival Probability Curves for Router Bits

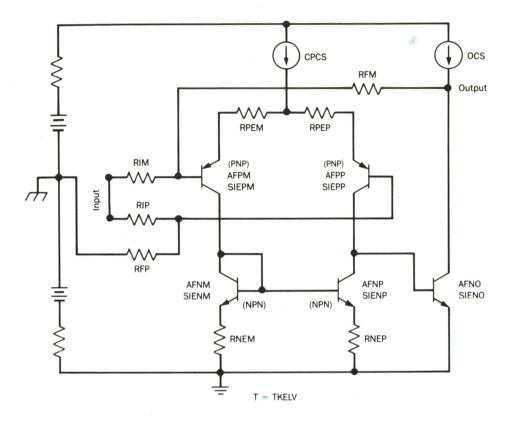

Figure 10.5. Circuit Diagram for a Differential Operational Amplifier

10.4.1 CONTROL AND NOISE FACTORS

The balancing property of the circuit dictates the following relationship among the nominal values of the various circuit parameters: RFP = RFM, RPEP = RPEM, RNEP = RNEM, AFPP = AFPM, AFNP = AFNM, SIEPP = SIEPM, and SIENP = SIENM. The circuit parameter names beginning with AF refer to the alpha parameter of the transistors and those beginning with SIE refer to the saturation currents of the transistors. Further, the gain requirements of the circuit dictate the following ratios of resistance values RIM = RFM/3.55 and RIP = RFM/3.55. These relationships among the circuit parameters are called tracking relationships.

There are only five control factors for this circuit: RFM, RPEM, RNEM, CPCS, and OCS. The transistor parameters could not be specified for this design because the manufacturing technology was preselected, and it dictated the nominal values of all transistor parameters. Also, the tracking relationships determine the nominal values of the remaining resistors.

The number of noise parameters is 21. They are the tolerances on the eight resistors, 10 transistor parameters corresponding to the five transistors, two current sources and the temperature. The mean values and tolerances for these noise factors are given in Table 10.3. The mean values of only the first five parameters are determined by the circuit design.

TABLE 10.3. NOISE FACTORS FOR THE DIFFERENTIAL OP-AMP CIRCUIT

Name	Mean	Tolerance	Levels (multiply by mean)		
			1	2	3
1. RFM	71 kilohms	1%	0.9967	1.0033	
2. RPEM	15 kilohms	21%	0.93	1.07	
3. RNEM	2.5 kilohms	21%	0.93	1.07	
4. CPCS	20 μA	6%	0.98	1.02	
5. OCS	20 μA	6%	0.98	1.02	
6. RFP	RFM	2%	0.9933	1.0067	
7. RIM	RFM/3.55	2%	0.9933	1.0067	
8. RIP	RFM/3.55	2%	0.9933	1.0067	
9. RPEP	RPEM	2%	0.9933	1.0067	
10. RNEP	RNEM	2%	0.9933	1.0067	
11. AFPM	0.9817	2.5%	0.99	1	1.01
12. AFPP	AFPM	1/2%	0.998	1	1.002
13. AFNM	0.971	2.5%	0.99	1	1.01
14. AFNP	AFNM	1/2%	0.998	1	1.002
15. AFNO	0.975	1%	0.99	1	1.01
16. SIEPM	3.OE-13A	Factor of 7	0.45	1	2.21
17. SIEPP	SIEPM	Factor of 1.214	0.92	1	1.08
18. SIENM	6.OE-13A	Factor of 7	0.45	1	2.21
19. SIENP	SIENM	Factor of 1.214	0.92	1	1.08
20. SIENO	6.OE-13A	Factor of 2.64	0.67	1	1.49
21. TKELV	298K	15%	0.94	1	1.06

The tolerances listed are the 3σ limits. Thus for RPEM the σ is $21/3 = 7$ percent of its nominal value. Further, the interdependency or the correlation among the noise factors is also expressed through tracking. To see this point, let us look at the relationship between RPEP and RPEM. The 21 percent tolerance for RPEM represents the specimen-to-specimen variation within a lot and the lot-to-lot variation. But on a given specimen, the two resistors RPEM and RPEP are located physically close together. So there is less variation between the two resistances. This is the origin of the correlation between the two resistances. Suppose in a particular specimen RPEM = 15 kilohms. Then for that specimen RPEP will vary around 15 kilohms with 3σ limits equal to 2 percent of 15 kilohms. If for another specimen RPEM is 16.5 kilohms (10 percent more than 15 kilohms), then RPEP will vary around 16.5 kilohms with 3σ limits equal to 2 percent of 16.5 kilohms.

The saturation currents are known to have long-tailed distributions. So the tolerances are expressed as multiplicative instead of additive. In other words, these tolerances are taken to be additive in the log domain.

10.4.2 EVALUATION OF MEAN SQUARED OFFSET VOLTAGE

For a particular design, that is, for a particular selection of the control factor values, the mean squared offset voltage can be evaluated in many ways. Two common methods are:

- Monte Carlo simulation. Random number generators are used to determine a large number of combinations of noise factor values. The offset voltage is evaluated for each combination and then the mean squared offset voltage is calculated. For obtaining accurate mean squared values, the Monte Carlo method usually needs a large number of evaluations of the offset voltage, which can be expensive.

- Taylor series expansion. In this method, one finds the first derivative of the offset voltage with respect to each noise factor at the nominal design point. Let $x_1, ..., x_k$ be the noise factors, with variances $\sigma_1^2, ..., \sigma_k^2$, respectively. Let v be the offset voltage. Then the estimated mean square offset voltage is

$$ r = \sum_{i=1}^{k} \left[\frac{\partial v}{\partial x_i} \right] \sigma_i^2. $$

Second-order Taylor series expansion is sometimes taken if curvatures and correlations are important. When the tolerances are large so that the non-linearities of v are important, the Taylor series approach does not give very accurate results.

In this application, however, we used the approach suggested by Taguchi. The orthogonal array, L_{36}, taken from Taguchi and Wu,[5] was used to estimate the mean squared offset voltage as a standardized measure to be optimized. Simulation studies reported by Taguchi during his trips to AT&T Bell Laboratories have shown that the orthogonal array method gives more precise estimates of variances and means when compared to the Taylor series expansion method.

For the resistance and current source tolerances two levels were chosen, situated one standard deviation on either side of the mean. These noise factors were assigned to columns 1 through 10 of the L_{36} matrix (see Table 10.4). For the 10 transistor parameters and the temperature, three levels were chosen, situated at the mean and at $\sqrt{3/2}$ times the standard deviation on either side of the mean. These noise factors were assigned to columns 12 through 22 of the matrix L_{36}. The submatrix of L_{36} formed by columns 1 through 10 and 12 through 22 is denoted by $\{J_{jl}\}$ and is referred to as the noise orthogonal array.

Each row of the noise array represents one specimen of differential op-amplifier with different values for the circuit parameters in accordance with the tolerances. Let v_j be the offset voltage corresponding to row j of the noise orthogonal array. Then the mean square offset voltage is estimated by

$$r = \frac{1}{36} \sum_{j=1}^{36} v_j^2 .$$

Since the most desired value of the offset voltage is zero, the appropriate S/N ratio to be maximized for optimizing this circuit is

$$\eta = -10 \log_{10} r .$$

TABLE 10.4. L_{36} ORTHOGONAL ARRAY

No.	1	2	3	4	5	6	7	8	9	10	11	12	13	14	15	16	17	18	19	20	21	22	23
1	1	1	1	1	1	1	1	1	1	1	1	1	1	1	1	1	1	1	1	1	1	1	1
2	1	1	1	1	1	1	1	1	1	1	1	2	2	2	2	2	2	2	2	2	2	2	2
3	1	1	1	1	1	1	1	1	1	1	1	3	3	3	3	3	3	3	3	3	3	3	3
4	1	1	1	1	1	2	2	2	2	2	2	1	1	1	1	2	2	2	2	3	3	3	3
5	1	1	1	1	1	2	2	2	2	2	2	2	2	2	2	3	3	3	3	1	1	1	1
6	1	1	1	1	1	2	2	2	2	2	2	3	3	3	3	1	1	1	1	2	2	2	2
7	1	1	2	2	2	1	1	1	2	2	2	1	1	2	3	1	2	3	3	1	2	2	3
8	1	1	2	2	2	1	1	1	2	2	2	2	2	3	1	2	3	1	1	2	3	3	1
9	1	1	2	2	2	1	1	1	2	2	2	3	3	1	2	3	1	2	2	3	1	1	2
10	1	2	1	2	2	1	2	2	1	2	1	1	3	2	1	3	2	3	2	1	3	2	1
11	1	2	1	2	2	1	2	2	1	2	1	2	1	3	2	1	3	1	3	2	1	3	2
12	1	2	1	2	2	1	2	2	1	2	1	3	2	1	3	2	1	2	1	3	2	1	3
13	1	2	2	1	2	2	1	2	1	2	1	1	2	3	1	3	2	1	3	3	2	1	2
14	1	2	2	1	2	2	1	2	1	2	1	2	3	1	2	1	3	2	1	1	3	2	3
15	1	2	2	1	2	2	1	2	1	2	1	3	1	2	3	2	1	3	2	2	1	3	1
16	1	2	2	2	1	2	2	1	2	1	1	1	2	3	2	3	2	1	1	3	1	3	2
17	1	2	2	2	1	2	2	1	2	1	1	2	3	1	3	1	3	2	2	1	2	1	3
18	1	2	2	2	1	2	2	1	2	1	1	3	1	2	1	2	1	3	3	2	3	2	1
19	2	1	2	2	1	1	2	2	1	2	1	1	2	1	3	3	3	1	2	2	1	2	3
20	2	1	2	2	1	1	2	2	1	2	1	2	3	2	1	1	1	2	3	3	2	3	1
21	2	1	2	2	1	1	2	2	1	2	1	3	1	3	2	2	2	3	1	1	3	1	2
22	2	1	2	1	2	2	2	1	1	1	1	2	1	2	2	3	3	1	2	1	1	3	2
23	2	1	2	1	2	2	2	1	1	1	1	2	2	3	3	1	1	2	3	2	2	1	3
24	2	1	2	1	2	2	2	1	1	1	1	2	3	1	1	2	2	3	1	3	3	2	1
25	2	1	1	2	2	2	1	2	2	1	1	1	3	2	1	2	3	3	1	3	1	2	2
26	2	1	1	2	2	2	1	2	2	1	1	2	1	3	2	3	1	1	2	1	2	3	3
27	2	1	1	2	2	2	1	2	2	1	1	3	2	1	3	1	2	2	3	2	3	1	1
28	2	2	2	1	1	1	1	2	2	1	2	1	3	2	2	2	1	1	3	2	3	1	3
29	2	2	2	1	1	1	1	2	2	1	2	2	1	3	3	3	2	2	1	3	1	2	1
30	2	2	2	1	1	1	1	2	2	1	2	3	2	1	1	1	3	3	2	1	2	3	2
31	2	2	1	2	1	2	1	1	1	2	2	1	3	3	3	2	3	2	2	1	2	1	1
32	2	2	1	2	1	2	1	1	1	2	2	2	1	1	1	3	1	3	3	2	3	2	2
33	2	2	1	2	1	2	1	1	1	2	2	3	2	2	2	1	2	1	1	3	1	3	3
34	2	2	1	1	2	1	2	1	2	1	2	1	3	1	2	3	2	3	1	2	2	3	1
35	2	2	1	1	2	1	2	1	2	1	2	2	1	2	3	1	3	1	2	3	3	1	2
36	2	2	1	1	2	1	2	1	2	1	2	3	2	3	1	2	1	2	3	1	1	2	3

10.4.3 OPTIMIZATION OF THE DESIGN

Orthogonal array experimentation is also an efficient way to maximize a non-linear function—in this case the maximization of η with respect to the control factors. The control factors and their alternate levels are listed in Table 10.5. The L_{36} array was used to simultaneously study the five control factors. (In this case, the array L_{18} would have been sufficient). The factors RFM, RPEM, RNEM, CPCS, and OCS were assigned to columns 12, 13, 14, 15, and 16, respectively. The submatrix of L_{36} formed by columns 12 through 16 is denoted by $\{I_{ik}\}$ and is referred to as the control orthogonal array.

TABLE 10.5. CONTROL FACTORS FOR THE DIFFERENTIAL OP-AMP CIRCUIT

Label	Name	Description	Levels		
			1	2	3
A	RFM	Feedback resistance, minus terminal (kilohms)	33.5	71	142
B	RPEM	Emitter resistance, PNP, minus terminal (kilohms)	7.5	15	30
C	RNEM	Emitter resistance, NPN, minus terminal (kilohms)	1.25	2.5	5
D	CPCS	Complementary pair current source (μA)	10	20	40
E	OCS	Output current source (μA)	10	20	40

Starting design

Each row of the control orthogonal array represents a different design. For each design the S/N ratio was evaluated using the procedure described under "Evaluation of Mean Squared Offset Voltage." The simulation algorithm is graphically displayed in Figure 10.6.

Standard analysis of variance was performed on the η values to generate Table 10.6. The effect of each factor on η is displayed in Figure 10.7. From Table 10.6 it is apparent that only RPEM, CPCS, and OCS have an effect on η that is much bigger than the error variance. The effect of RPEM is the largest, and there is indication that reduction in its value below 7.5 kilohms could give even more improvement in offset voltage. For both current sources, 10 μA to 20 μA seems to be the flat region, indicating that we are very near the best values for these parameters. Also, the potential improvement by changing these current sources from 10 μA to 20 μA seems small. Thus we chose the following two designs as potential optimum points:

- Optimum 1: Only change RPEM from 15 kilohms to 7.5 kilohms. By the procedure in "Evaluation of Mean Squared Offset Voltage," the value of η for this design was found to be 33.70 dB compared to 29.39 dB for the starting design. In terms of the rms offset voltage this represents an improvement from 33.9 mV to 20.7 mV.

- Optimum 2: Change RPEM to 7.5 kilohms. Also change both CPCS and OCS to 10 μA. The η for this design was computed to be 35.82 dB, and rms offset voltage was seen to be 16.2 mV.

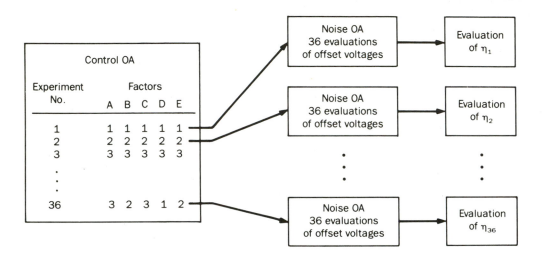

Figure 10.6. Simulation Algorithm

TABLE 10.6. ANALYSIS OF VARIANCE FOR η = −10 log₁₀
(MEAN SQUARED OFFSET VOLTAGE)

Control factor	Level means			Sum of squares	Degrees of freedom	Mean square	F
	1	2	3				
A. RFM	26.5	26.4	25.3	9.9	2	4.95	0.5
B. RPEM	30.3	26.4	21.5	463.7	2	231.85	25.0
C. RNEM	25.1	25.8	27.3	29.9	2	14.95	1.6
D. CPCS	27.5	27.1	23.6	111.1	2	55.55	6.0
E. OCS	27.3	27.0	23.8	87.5	2	43.75	4.7
Error				231.6	25	9.26	

Overall mean = 26.05

In the discussion so far, we have paid attention to only the dc offset voltage. Stability under ac operation is also an important consideration. For a more elaborate study of this characteristic, one must generate more data like those in Table 10.6 and Figure 10.7. The optimum control factor setting should then be obtained by jointly considering the effects on both the dc and ac characteristics. If conflicts occur, appropriate trade-offs can be made using the quantitative knowledge of the effects. In our study, we simply checked for ac stability at the two optimum conditions. For sufficient safety margin with respect to ac stability, we selected optimum 1 as the best design and called it simply the optimum design.

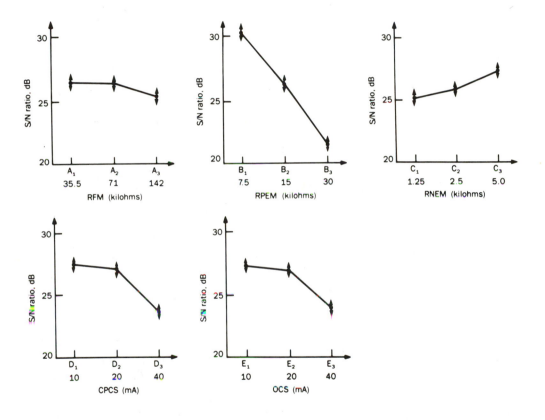

Figure 10.7. Average Effects of Control Factors on $\eta = -10 \log_{10}$ **(Mean Squared Offset Voltage).** The 2 σ limits are indicated by the arrows.

10.4.4 TOLERANCE DESIGN

With the sensitivity to noise minimized for the dc offset voltage, the next step is to examine the contribution to the mean squared offset voltage by each noise source. By performing analysis of variance of the 36 offset voltages corresponding to the noise orthogonal array for the optimum control factor settings, we obtained Table 10.7, which gives the breakdown of the contribution of each tolerance to the total mean squared offset voltage. The table also gives a similar breakdown of the mean squared offset voltage for the starting design. The reduction in the sensitivity of the design to various noise sources is apparent.

TABLE 10.7. BREAKDOWN OF MEAN SQUARE OFFSET VOLTAGE BY NOISE SOURCES

	Contribution to mean square offset voltage $(10^{-4}\ V^2)$	
Noise factor*	Starting Design	Change RPEM to 7.5 kilohms
SIENP	6.7	2.3
AFNO	2.2	0.8
AFNM	1.6	0.4
SIEPP	0.5	0.5
RPEP	0.2	0.1
AFPP	0.1	0.1
Remainder	0.2	0.1
Total	11.5	4.3

* The six largest contributors to the mean squared offset voltage are listed here.

The table exhibits the typical Pareto principle. That is, a small number of tolerances account for most of the variation in the response. In particular, SIENP has the largest contribution to the mean squared offset voltage. Hence it is a prime candidate for reducing the tolerance, should we wish to further reduce the mean squared offset voltage. AFNO, AFNM, and SIEPP have moderate contribution to the mean squared offset voltage. The rest of the noise factors contribute negligibly to the mean squared offset voltage. So if it will yield further manufacturing economies, relaxing these tolerances should be considered carefully.

10.4.5 REDUCING COMPUTATIONAL EFFORT

The optimization of the differential op-amp discussed earlier needed 36×36 = 1296 evaluations of the circuit response—namely the offset voltage. Although in this case the computer time needed was not an issue, in some cases making an evaluation of each response can be expensive. In such cases, significant reduction of the computational effort can be accomplished by using qualitative engineering knowledge about the effects of the noise factors. For example, we form a *composite noise factor*. The high level of the composite noise factor corresponds to the combination of the individual noise factor levels that give high response. The *low* and *nominal* levels of the composite noise factor can be defined similarly. Thus the size of the noise orthogonal array can be drastically reduced, leading to much smaller computational effort during design optimization. Further reduction in computational effort can be obtained by taking only two levels for each noise factor; however, generally we recommend three levels. For tolerance design, we need to identify the sensitivity of each noise factor. Therefore a composite noise factor would have to be dissolved into its various components.

10.5 CONCLUSION

The robust design method has now been used in many areas of engineering throughout the United States. For example, robust design has lead to improvement of several processes in *very large scale integration (VLSI) fabrication*: window photolithography, etching of aluminum lines, reactive ion etching processes, furnace operations for deposition of various materials, and so forth. These processes are used for manufacturing 1-megabit and 256-kilobit memory chips, 32-bit processor chips, and other products. The *window photolithography* application documented in Phadke and others[8] was the first application in the United States that demonstrated the power of Taguchi's approach to quality and cost improvement through robust process design. In particular, the benefits of the application were:

- Fourfold reduction in process variance.

- Threefold reduction in fatal defects.

- Twofold reduction in processing time. This resulted from the fact that the process became stable so that time-consuming inspection could be dropped.

- Easy transition of design from research to manufacturing.

- Easy adaptation of the process for finer line technology, which is usually a very difficult problem.

The *aluminum etching* application was an interesting one in that it originated from a belief that poor photoresist print quality leads to line width loss and to undercutting. By making the process insensitive to photoresist profile variation and other sources of variation, the visual defects were reduced from 80 percent to 15 percent. Moreover, the etching step could then tolerate the variation in photoresist profile.

In *reactive ion etching* of tantalum silicide, the process gave highly nonuniform etch quality, so only 12 out of 18 possible wafer positions could be used for production. After optimization, 17 wafer positions became usable—a hefty 40 percent increase in machine capacity. Also, the efficiency of the orthogonal array experimentation allowed this project to be completed in the 20-day deadline. Thus in this case $1.2 million was saved in equipment replacement costs not counting expense of disruption on the factory floor (Katz and Phadke[9]).

The *router bit life improvement* project described in this article led to a two- to fourfold increase in the life of router bits used in cutting printed wiring boards. The project illustrates how reliability or life improvement projects can be organized to find best settings of control factors with a very small number of samples. The number of samples needed in this approach is very small yet it can give valuable information about how each factor changes the survival probability curve.

In the *differential op-amp circuit optimization* example described in this article, a 40-percent reduction in the rms offset voltage was realized by simply finding a new design center. This was done by reducing sensitivity to all tolerances and temperature, rather than reducing tolerances, which could have increased manufacturing cost.

Here the noise orthogonal array was used in a novel way—to efficiently simulate the effect of many noise factors. This approach can be beneficially used for evaluating designs and for system or software testing. Further, the approach can be automated and made to work with various computer-aided design tools to make design optimization a routine practice.

This approach was also used to find optimum proportions of ingredients for making *water-soluble flux* (Lin and Kackar[10]). By simultaneous study of the parameters for the wave soldering process and the flux composition, the defect rate was reduced by 30 to 40 percent.

Orthogonal array experiments can be used to tune hardware/software systems (Pao, Phadke, and Sherrerd[11]). By simultaneous study of three hardware and six software parameters, the response time of the UNIX[5] operating system was reduced 60 percent for a particular set of load conditions experienced by the machine.

Under the leadership of American Supplier Institute and Ford Motor Company, a number of automotive suppliers have achieved quality and cost improvement through robust design. Many of these applications are documented by American Supplier Institute.[12]

These examples show that robust design is a collection of tools and comprehensive procedures for simultaneously improving product quality, performance, and cost, and also engineering productivity. Its widespread use in industry is bound to have a far-reaching economic impact.

ACKNOWLEDGMENTS

The two applications described in this paper would not have been possible without the collaboration and diligent efforts of many others. The router bit life

---

5. Registered trademark of AT&T.

improvement study was done in collaboration with Dave Chrisman of AT&T Technologies, Richmond Works. The optimization of the differential op-amp was done jointly with Gary Blaine, a former member of the technical staff at AT&T Bell Laboratories, and Joe Leanza of AT&T Information Systems. The survival probability curves were obtained with the software package for analysis of variance developed by Chris Sherrerd of Bell Laboratories. Rajiv Keny and Paul Sherry of Bell Laboratories provided helpful comments on this paper.

REFERENCES

1. Taguchi, G. 1978. Off-Line and On-Line Quality Control Systems. *International Conference on Quality Control* (Tokyo).

2. Taguchi, G., and M. S. Phadke. 1984. Quality Engineering through Design Optimization. *Conference Record, GLOBECOM 84 Meeting*, 1106-1113. IEEE Communications Society, Atlanta (Nov).
(Quality Control, Robust Design, and the Taguchi Method; Article 5)

3. Phadke, M. S. 1982. Quality Engineering Using Design of Experiments. *Proceedings of the American Statistical Association* Section on Statistical Education, 11-20. Cincinnati. (Aug.)
(Quality Control, Robust Design, and the Taguchi Method; Article 3)

4. Kackar, R. N. 1985. Off-line Quality Control, Parameter Design, and the Taguchi Method. *Journal of Quality Technology* (Oct.):176-188.
(Quality Control, Robust Design, and the Taguchi Method; Article 4)

5. Taguchi, G., and Yu-In Wu. 1979. *Introduction to Off-Line Quality Control.* Nagaya, Japan: Central Japan Quality Control Association, Meieki Nakamura-Ku. Available in English through the American Supplier Institute, Inc.

6. Hicks, C. R. 1973. *Fundamental Concepts in the Design of Experiments.* New York: Holt, Rinehart and Winston.

7. Box, G. E. P., W. G. Hunter, and I. S. Hunter. 1978. *Statistics for Experimenters—An Introduction to Design, Data Analysis and Model Building.* New York: John Wiley and Sons, Inc.

8. Phadke, M. S., R. N. Kackar, D. V. Speeney, and M. J. Grieco. 1983. Off-Line Quality Control in Integrated Circuit Fabrication Using Experimental Design. *The Bell System Technical Journal* **1** (May-June, no. 5):1273-1309.
(Quality Control, Robust Design, and the Taguchi Method; Article 6)

9. Katz, L. E., and M. S. Phadke. 1985. Macro-Quality with Micro-Money. *AT&T Bell Laboratories Record* (Nov):22-28.

(Quality Control, Robust Design, and the Taguchi Method; Article 2)

10. Lin, K. M., and R. N. Kackar. n.d. Wave Soldering Process Optimization by Orthogonal Array Design Method. *Electronic Packaging and Production*. Forthcoming.

11. Pao, T. W., M. S. Phadke, and C. S. Sherrerd. 1985. Computer Response Time Using Orthogonal Array Experiments. *Conference Record* vol. 2, 890-895. IEEE International Communications Conference, Chicago (June 23-26).

12. American Supplier Institute, Inc. 1984-1985. *Proceedings of Supplier Symposia on Taguchi Methods*. Romulus, Mich.

PART THREE

METHODOLOGY

11

TESTING IN INDUSTRIAL EXPERIMENTS WITH ORDERED CATEGORICAL DATA

Vijayan N. Nair

11.1 INTRODUCTION

The resurgence in the use of experimental design methods in industrial applications in the past few years is primarily due to the pioneering work of Professor G. Taguchi. Traditional experimental design techniques focus on identifying factors that affect the level of a production or manufacturing process. We call this the location effects of the factors. Taguchi was the first to recognize that statistically planned experiments could and should be used in the product development stage to detect factors that affect the variability of the output. This will be termed the dispersion effects of the factors. By setting the factors with important dispersion effects at their "optimal" levels, the output can be made robust to changes in operating and environmental conditions in the production line. Thus the identification of dispersion effects is also crucial in improving the quality of a process.

This article deals with techniques for analyzing ordered categorical data from industrial experiments for quality improvement. The method recommended by Taguchi[1, 2] and currently used in industry for analyzing such data is a technique called *accumulation analysis*. Taguchi's accumulation analysis statistic is shown to have reasonable power for identifying the important location effects; however, it is an unnecessarily complicated procedure for just identifying the location effects. Moreover, it performs poorly in determining the important dispersion effects. It also confounds the two effects, so that one cannot separate the factors that affect variability from those that affect location. Instead two sets of scores are proposed for identifying the location and dispersion effects separately. They are intended as simple, approximate procedures for determining the important factors.

The article is organized as follows. It begins with an overview of the parameter design problem in off-line quality control in Section 11.2. This provides a motivation for estimating dispersion effects. Since the statistical literature on Taguchi's accumulation analysis is not easily available, we review the technique and suggest some modifications in Section 11.3. Its application in an experiment to optimize the process of forming contact windows in semiconductor circuits (Phadke and others[3]) at AT&T Bell Laboratories is also described. Some of the properties of the technique are given in Section 11.4. These are based on the decomposition of the statistic into orthogonal components in Nair.[4] It is shown that the first two components of the accumulation analysis statistic test primarily for differences in location and dispersion respectively. In fact, in a special case, the first component is Wilcoxon's test for location and the second component is Mood's[5] test for dispersion, both applied to contingency tables. Taguchi's statistic gives most weight to the first component and hence is good, primarily, for detecting location effects. For detecting dispersion effects, however, it may not even be as powerful as Pearson's chi-square statistic which gives equal weight to all the components. I propose instead the use of some simple scoring methods in Section 11.5. Scoring methods are particularly easy to use in analyzing multidimensional contingency table data. They are illustrated by reanalyzing Phadke and others[3] contact window data.

In quality improvement experiments, there is typically little interest in detailed structural modeling and analysis of the data. The user's primary goal is to identify the really important factors and determine the levels to improve process quality. For these reasons, users may find the simple scoring schemes proposed in this paper to be attractive. It should be noted, however, that there are a variety of other methods available in the statistical literature for analyzing ordered categorical data (see Agresti[6]). For additional literature on recent research in experimental design for quality improvement, see Box;[7] Kackar;[8] León, Shoemaker, and Kackar[9] Taguchi;[10] and Taguchi and Wu.[11]

11.2 OVERVIEW OF PARAMETER DESIGN PROBLEM

In this section, I present a statistical formulation of the parameter design problem in off-line quality control. I consider only the fixed target case. See also Taguchi and Wu,[11] and Léon, Shoemaker and Kackar.[9]

We are interested in designing a process whose output Y is a function of two sets of factors: (a) noise factors[1] (all of the uncontrollable factors including the

---

1. Glossary equivalent is *noise parameter*.

operating and environmental conditions) and (b) design factors (factors that can be controlled and manipulated by the process engineer). Let $\mathbf{n}$ and $\mathbf{d}$ denote the settings of the noise and design factors respectively. The target value for this process is fixed at t_0, and the cost when the output Y deviates from the target t_0 is measured by the loss function $L(Y, t_0)$. The goal of the experiment is to determine the design parameters—that is, settings of the design factors $\mathbf{d}$ to minimize the average loss $E_\mathbf{n}[L(Y, t_0)] = E_\mathbf{n}[L(f(\mathbf{d}, \mathbf{n}), t_0)]$.

In most cases, the data $Y = f(\mathbf{d}, \mathbf{n})$ can be transformed to, approximately, a location-scale model so that

$$\tau(Y) \equiv Z \approx \mu(\mathbf{d}) + \sigma(\mathbf{d})\,\varepsilon \tag{1}$$

where the stochastic component ε does not depend on the settings of design factors $\mathbf{d}$, and μ and σ measure, respectively, the location and dispersion effects of $\mathbf{d}$. Assume without loss of generality that ε in (1) has mean 0 and variance 1. Suppose that the loss function can be approximated by squared error loss in terms of the transformed variable; that is,

$$L(Y, t_0) = [Z - \tau(t_0)]^2. \tag{2}$$

Then to minimize the average loss, we need to choose the design factors to minimize the mean squared error

$$E_\mathbf{n}\,L(Y, t_0) = E[Z - \tau(t_0)]^2$$

$$= [\mu(\mathbf{d}) - \tau(t_0)]^2 + \sigma^2(\mathbf{d}). \tag{3}$$

That is, the goal of the experiment is to identify the important design factors that affect μ and those that affect σ and set them at the "optimal" levels (within the operating constraints) to minimize the mean squared error (3). This is in contrast with traditional experimental design methods that focus merely on identifying the location effects. They treat $\sigma(\mathbf{d})$ as nuisance parameters and usually assume that they are constant (no dispersion effects) or are known up to a scale parameter as $\mathbf{d}$ varies.

As noted by León, Shoemaker and Kackar,[9] and Taguchi and Wu,[11] in many situations there exist readily identifiable factors, known as "adjustment" factors,[2] that mainly affect the level (location) of a process but not its dispersion. In

---

2. Glossary equivalent is *adjustment parameter*.

such cases, we need to minimize only the variance in (3), since the bias component can be made small by fine-tuning the adjustment factors. So the location effects play the role of nuisance parameters and the primary emphasis is in determining the dispersion effects.

When the output Y from the industrial experiment is a continuous variable, a reasonable way of analyzing the data is: (a) carry out the usual ANOVA on $Z_i = \tau(Y_i)$ to identify location effects, and (b) get an estimate $\hat{\sigma}_i$ at each experimental run and do an ANOVA on $\log \hat{\sigma}_i$ at each experimental to identify dispersion effects. Taguchi recommends analyzing log of the signal-to-noise ratio (inverse of the coefficient of variation). This yields essentially the same results as (b) above when $Z_i = \log(Y_i)$, perhaps the most common transformation in industrial applications. The analysis in (a) is inefficient since it ignores the fact that the variances are unequal. However, it will be adequate in most cases. A more formal analysis can be done by fitting a parametric model to (1) and using maximum likelihood techniques (see Hartley and Jayatillake[12]).

11.3 ACCUMULATION ANALYSIS: THE METHOD AND AN EXAMPLE

In some industrial experiments, the output consists of categorical data with an ordering in the categories. For analyzing such data, Taguchi[1, 2] proposed the accumulation analysis method as an alternative to Pearson's chi-square test. His motivation for recommending this technique appears to be its similarity to ANOVA for quantitative variables. The use of ANOVA techniques with dichotomous data has been discussed in the literature as early as Cochran.[13] See also Pitman[14] for randomization tests valid for samples from any population. More recently, Light and Margolin[15] proposed a method called CATANOVA by defining an appropriate measure of variation for categorical data. Unlike these methods, however, Taguchi considers situations with ordered categories and does ANOVA on the cumulative frequencies.

To define the method, consider a one-factor experiment with factor A at I levels. For simplicity of exposition, assume there are an equal number, n, of observations at each level. The observations are classified into one of K ordered categories, and Y_{ik} denotes the observed frequency in category k at level i ($k = 1, ..., K; i = 1, ..., I$). Denote the cumulative frequencies by $C_{ik} = \sum_{j=1}^{k} Y_{ij}$ and by their averages across levels $C_{.k} = \sum_{i=1}^{I} C_{ik}/I$. Then the sum of squares (SS) for factor A is given by

$$SS_A = n \sum_{k=1}^{K-1} \sum_{i=1}^{I} (C_{ik} - C_{.k})^2 / [C_{.k}(n - C_{.k})]. \tag{4}$$

This is obtained as follows. From the cumulative frequencies in the kth column, we get the sum of squares for factor A as $SS_{A,k} = \sum_{i=1}^{I}(C_{ik} - C_{.k})^2$. Since these have different expectations under the null hypothesis that factor A has no effect, they are standardized before being combined to get the single sum of squares in (4). As a crude approximation, Taguchi[2] suggests using $(I-1)(K-1)$ degrees of freedom (df) for SS_A.

Since the cumulative frequency C_{ik} in the (i, k)th cell is made up of C_{ik} ones and $(n - C_{ik})$ zeros, the within sum of squares is proportional to $[C_{ik}(n - C_{ik})]$. By combining these, Taguchi[2] proposed the sum of squares for error

$$SS_e = n \sum_{k=1}^{K-1} \sum_{i=1}^{I} [C_{ik}(n - C_{ik})] / [C_{.k}(n - C_{.k})]. \tag{5}$$

The expectation of SS_e is $[n(n-1)I^2(K-1)/(In-1)] \approx I(n-1)(K-1)$. Perhaps because of this, Taguchi[2] suggests taking $I(n-1)(K-1)$ as df for SS_e (!) and using the statistic

$$F_A = MS_A/MS_e \tag{6}$$

to test for the effect of factor A (MS denotes mean square). The preceding definitions extend in a straightforward way to multifactor situations.

The use of MS_e in the denominator of (6) is unnecessary. There is really no notion of "error" in this situation since SS_A in (4) has already been standardized by $C_{.k}(1 - C_{.k}/n)$, the conditional variance of C_{ik} given the column margins. Further, we see from the last paragraph that, under the null hypothesis of no effect, the expected value of MS_e is approximately 1. So there is nothing to be gained from using MS_e. When one or more of the factors have large effects, MS_e can be substantially smaller than 1, thus inflating the F statistic.

From now on, I shall use the terms accumulation analysis and Taguchi's statistic to mean just the sum of squares in the numerator SS_A, SS_B, and so forth. The generic notation T will be used to denote the statistics. As observed by Taguchi and Hirotsu,[16] $T = \sum_{k=1}^{K-1} P_k$ where P_k is the Pearson's chi-square statistic based on categories 1 through k versus categories $(k+1)$ through K. For this reason, T is also sometimes referred to as the cumulative chi-squared statistic.

We now consider an example. Phadke and others[3] conducted an experiment to optimize the process of forming contact windows in complementary metal-oxide semiconductor circuits. Contact windows facilitate interconnections between the gates, sources and drains in circuits. The target size for the windows was about 3.0 μm. It was important to produce windows near the target dimension; windows not open or too small result in loss of contact to the devices, and excessively large windows lead to shorted device features.

The process of forming windows involves photolithography. Phadke and others[3] identified 8 factors that were important in controlling window size: *A*—mask dimension, *BD*—viscosity × bake temperature, *C*—spin speed, *E*—bake time, *F*—aperture, *G*—exposure time, *H*—developing time, and *I*—plasma etch time. The experimental design was an L_{18} orthogonal array (see Taguchi and Wu[11]) with factor *A* at two levels and all others at 3 levels. There were 10 observations at each experimental run. One of the measurements made from the experiment was post-etch window size of test windows. Many of the windows were not open, and window sizes could be measured only from those that were open. Phadke and others[3] grouped the data into five categories (see Table 11.1) and used the accumulation analysis method to analyze the data.

TABLE 11.1. POST-ETCH WINDOW SIZE DATA—FREQUENCIES

Expt. No.	I	II	III	IV	V
1	10	0	0	0	0
2	0	3	3	2	2
3	1	0	0	9	0
4	10	0	0	0	0
5	10	0	0	0	0
6	5	3	2	0	0
7	10	0	0	0	0
8	5	0	0	5	0
9	0	1	4	5	0
10	2	5	3	0	0
11	1	1	2	6	0
12	1	0	1	3	5
13	5	0	3	2	0
14	6	3	1	0	0
15	10	0	0	0	0
16	10	0	0	0	0
17	0	0	4	3	3
18	0	0	0	0	10
Overall proportion	.48	.09	.13	.19	.11

I: Window not open   IV: [2.75, 3.25] µm

II: (0,2.25) µm   V: 3.25 µm

III: [2.25, 2.75) µm

The ANOVA results from accumulation analysis are given in Table 11.2. It differs from Phadke and others[3] ANOVA table in several ways. They decomposed the error sum of squares into further components based on a random effects analysis and considered the F-statistics given by (6). I consider the MS's of the factors directly and provide the error SS only to verify the point made earlier that when there are many significant factors, the error MS is rather small. The df's in Table 11.2 are inappropriate and a more reasonable approximation is discussed in the next section. This is not a serious issue, however, since as emphasized by Taguchi, these statistics are not to be used for formal tests of hypotheses but more as indicators of the importance of the different factors.

TABLE 11.2. ANOVA FROM ACCUMULATION ANALYSIS —POST-ETCH WINDOW SIZE DATA

Source	df	SS	MS
A	4	26.64	6.66
BD	8	112.31	14.04
C	8	125.52	15.69
E	8	36.96	4.62
F	8	27.88	3.49
G	8	42.28	5.29
H	8	45.57	5.70
I	8	23.80	2.98
Error	656	279.04	0.43
Total	716	720.00	

When Table 11.2 is viewed thus, we can conclude that viscosity-bake temperature (BD) and spin speed (C) are the two most important factors affecting window sizes. Mask dimension (A), exposure time (G), and developing time (H) are important to a lesser extent. Phadke and others[3] arrived at the same conclusions from their analysis. It is interesting to compare these conclusions with those obtained from the classical Pearson's chi-squared tests. The SS's in Table 11.3 are the values of Pearson's chi-square statistics for testing the homogeneity hypothesis for each factor based on the marginal tables. Interpreting the MS's as before, we see that viscosity-bake temperature (BD) and spin speed (C) are again the most important factors. They are not as dominant here, however, compared to Table 11.2. Bake time (E) is an important factor in Table 11.3, something that was not observed from accumulation analysis.

The results in the next two sections explain the differences between the two methods of analysis. It is shown that Taguchi's statistic is geared primarily towards determining the important location effects. In this regard, factors BD and C are indeed dominant. Factor E shows up as important in Table 11.3 because it has a large dispersion effect. In fact, as seen in Section 11.5, the factor that is most important in controlling variability of window sizes is etch time (I).

TABLE 11.3. PEARSON'S CHI-SQUARE TESTS FOR THE MARGINAL TABLES—POST-ETCH WINDOW SIZE DATA

Source	df	SS	MS
A	4	18.51	4.63
BD	8	51.01	6.38
C	8	49.59	6.20
E	8	38.20	4.78
F	8	27.07	3.38
G	8	43.81	5.48
H	8	34.30	4.29
I	8	26.21	3.28

11.4 ACCUMULATION ANALYSIS: SOME PROPERTIES

In this section, some interesting properties of T and its relationship to Pearson's chi-square statistic P are presented. They are based on the decomposition of the statistics into orthogonal components in Nair.[4] Some power comparisons of T and P are also given.

Consider again the one-factor situation described at the beginning of Section 11.3. T is given by (4). Pearson's statistic for testing for the effect of factor A is

$$P = \sum_{k=1}^{K} \sum_{i=1}^{I} (Y_{ik} - Y_{.k})^2 / Y_{.k}. \tag{7}$$

Consider the $I \times K$ table with the column totals fixed. Let $\mathbf{R}$ denote the conditional variance-covariance matrix of $\mathbf{y}_i = (Y_{i1}, ..., Y_{iK})$, which is the same for $i = 1, ..., I$ under the null hypothesis. The statistic T can be expressed as a quadratic form $\sum_{i=1}^{I}(\mathbf{y}_i' \mathbf{A}' \mathbf{A} \mathbf{y}_i)$ for some matrix $\mathbf{A}$. The matrix $\mathbf{A}'\mathbf{A}$ can be decomposed as

$$\mathbf{A}'\mathbf{A} = \mathbf{Q} \wedge \mathbf{Q}' \tag{8}$$

where $\wedge$ is a $(K-1) \times (K-1)$ matrix of eigenvalues, $\mathbf{Q} = [\mathbf{q}_1 \mid ... \mid \mathbf{q}_{K-1}]$ is a $K \times (K-1)$ matrix of eigenvectors, and $\overline{\mathbf{Q}} = [1 \mid \mathbf{Q}]$ has the property that $\overline{\mathbf{Q}}'\mathbf{R}\overline{\mathbf{Q}}$ is proportional to the identity matrix. If $Z_j^2 = \sum_{i=1}^{I}(\mathbf{q}_j' \mathbf{y}_i)^2$ and λ_j's are the eigenvalues, then T has the representation

$$T = \sum_{j=1}^{K-1} \lambda_j Z_j^2. \tag{9}$$

The statistic P can be expressed in terms of these orthogonal components Z_j^2 as

$$P = \sum_{j=1}^{K-1} Z_j^2 . \qquad (10)$$

When $I = 2$, the Z_j's are, conditionally and under the null hypothesis, uncorrelated with mean zero and variance $(1-1/N)^{-1}$ where N is the total number of observations. For $I \geq 2$, Z_j^2, $j = 1, ..., K - 1$, are approximately independently and identically distributed as χ_{I-1}^2 when the sample sizes are large. It is clear from (9) that, for large n, T is distributed as a linear combination of chi-squared random variables. By matching the first two moments, this distribution can be approximated by $d\chi_v^2$ where

$$d = \sum_{j=1}^{K-1} \lambda_j^2 \Big/ \sum_{j=1}^{K-1} \lambda_j = 1 + \frac{2}{K-1} \sum_{k=1}^{K-1} \sum_{l=k+1}^{K-1} [C_{.k} \ (n - C_{.l})] / [(n - C_{.k}) C_{.l}]$$

and

$$v = (I-1)(K-1)/d. \qquad (11)$$

Simulation results indicate that this provides a reasonable approximation to the critical values of T.

When the (conditional) total frequencies in each column are the same, the decomposition of T in (9) can be obtained explicitly (Nair[4]). The eigenvectors turn out to be related to Chebychev polynomials; that is, the jth eigenvector $\mathbf{q}_j$ is a jth degree polynomial in k, $k = 1, ..., K$, and the jth eigenvalue is

$$\lambda_j = K/[j(j+1)] . \qquad (12)$$

So when $I = 2$, Z_1 is proportional to

$$W = \sum_{k=1}^{K} k \ (Y_{1k} - Y_{.k}), \qquad (13)$$

the grouped data version of Wilcoxon's statistic (Lehmann[17]) and it is known that W has good power even in contingency tables for detecting differences in underlying location models. Z_2 is proportional to

$$M = \sum_{k=1}^{K} (k - \tfrac{1}{2} - K/2)^2 \ (Y_{1k} - Y_{.k}), \qquad (14)$$

223

the grouped data version of Mood's[5] statistic. Mood proposed his statistic to test for differences in scale assuming that there are no (or known) location differences. The modification of Mood's test for scale in the presence of unknown location differences, considered by Sukhatme,[18] involves first estimating the location parameters using the original data and applying the rank test to the residuals. This modification cannot be applied to contingency tables. The asymptotic calculations in the Appendix, as well as the simulation results later in this section and Section 11.5 show that the statistic in (14) can be used to test for scale differences even in the presence of moderate location effects. When the location differences are large, it is difficult to make inferences about dispersion effects from categorical data.

For $I > 2$, in the equiprobable case, Z_1^2 is proportional to

$$\sum_{i=1}^{I} \left[\sum_{k=1}^{K} k (Y_{ik} - Y_{\cdot k}) \right]^2$$

is the Kruskal-Wallis statistic for contingency tables (Lehmann[17]). Similarly, Z_2^2 is proportional to

$$\sum_{i=1}^{I} \left[\sum_{k=1}^{K} (k - \frac{1}{2} - K/2)^2 (Y_{ik} - Y_{\cdot k}) \right]^2 ,$$

the generalization of the two-sided version of (14). Note from (12) that $\sum_{j=1}^{K-1} \lambda_j = K - 1$, the same as the total weight given by P to the components. T puts most weight, $\lambda_1 = K/2$, on the first component Z_1^2 which is a test for location effects. Therefore, accumulation analysis in fact does have good power for detecting ordered alternatives induced by location shifts. However, it is a much more complicated procedure than Z_1^2 for detecting location differences. T gives only a weight of $\lambda_2 = K/6$ to the second component Z_2^2, which tests for dispersion effects. So T may not have even as much power as P, which gives a weight of 1 to Z_2^2. Both P and T will have relatively poor power compared to Z_2^2 itself in detecting dispersion effects.

In the general non-equiprobable case, the components Z_1^2 and Z_2^2 do not coincide with Wilcoxon's and Mood's tests, respectively. As the simulation results in Table 11.4 show, however, the first two components still have good power for detecting location and scale differences, respectively. The set up for the Monte Carlo simulation results in Table 11.4 was as follows. There were 10,000 simulations of a 2^2 factorial experiment with factors A and B. For each simulation, there were 10 observations at the ith experimental run ($i = 1, ..., 4$), and the observations fell into one of five categories. Let $\pi_{ik} = \sum_{j=1}^{k} p_{ij}$ denote the cumulative probabilities ($k = 1, ..., 5; i = 1, ..., 4$). I examined the power behavior under a logistic location-shift model and a logistic scale-shift model. Specifically, with $(p_{.1}, \ldots, p_{.5}) = (.1, .15, .3, .2, .25)$ and $\pi_{.k} = \sum_{j=1}^{k} p_{.j}$, under the location-shift model,

$$\text{logit } \pi_{ik} = \text{logit } \pi_{.k} \pm \gamma_A \pm \gamma_B \pm \gamma_{AB} \tag{15}$$

and under the scale-shift model,

$$\text{logit } \pi_{ik} = (\text{logit } \pi_{.k}) \exp(\pm \gamma_A \pm \gamma_B \pm \gamma_{AB}). \tag{16}$$

To obtain the critical values, I used the χ_4^2 approximation for P, the approximation in (11) for T, and the χ_1^2 approximation for Z_1^2 and Z_2^2.

TABLE 11.4. POWER COMPARISON OF P, T, Z_1^2 and $Z_2^2(p_{.1}, \ldots, (p_{.5}) = (.1, .15, .3, .2, .25), \alpha = 0.10$. The power computations are for $\gamma_A = .5$, $\gamma_B = .25$ and $\gamma_{AB} = 0$ in the location [scale] model in (4.9) [(4.10)].

| | Achieved Level | | | Power | | | | | |
| | | | | Location Model | | | Scale Model | | |
Statistics	γ_A	γ_B	γ_{AB}	γ_A	γ_B	γ_{AB}	γ_A	γ_B	γ_{AB}
P	.10	.09	.09	.34	.14	.10	.68	.20	.11
T	.09	.09	.09	.49	.19	.08	.59	.16	.09
Z_1^2	.10	.10	.10	.52	.21	.09	.12	.11	.11
Z_2^2	.10	.10	.10	.11	.10	.11	.85	.30	.07

From Table 11.4, we see that (11) provides a reasonable approximation to the critical values of T. The power calculations were computed for the location model and the scale model with $\gamma_A = .5$, $\gamma_B = .25$, and $\gamma_{AB} = 0$ (no interaction in [15] and [16], respectively). In the location-shift model, T is more powerful than P, because it gives most of its weight to the first component Z_1^2; the power of Z_2^2 in this case is close to the level .10. In the scale-shift model, however, T is not even as powerful as P. This is because the power of its dominant component Z_1^2 is close to the level .10. Z_2^2, on the other hand, is good for detecting the dispersion effects. Simulations with several other configurations of $(p_{.1}, \ldots, p_{.5})$ gave qualitatively similar results.

11.5 TWO SIMPLE SETS OF SCORES FOR DETECTING LOCATION AND DISPERSION EFFECTS

To actually compute the first two components of Taguchi's statistic, Z_1^2, and Z_2^2, one needs to solve an eigen-problem. In this section, I propose two sets of easily computable scores that yield statistics very close to Z_1^2 and Z_2^2 and have comparable power.

The motivation for considering these particular sets of scores comes from the decomposition of Taguchi's statistic in the special equiprobable case. Recall that in this case the first component is the Wilcoxon (or its generalization the Kruskal-Wallis) test applied to contingency tables. It is computed by treating all the observations in a certain category as being tied and assigning them a score equal to the midrank for that category. (See also Bross[19] for "ridit" analysis). Since Wilcoxon's statistic is known to have reasonably good properties (see, i.e., Lehmann[17]), one can take it to be the first component even in the general non-equiprobable case. For the second component, the scores are taken to be quadratic in the midranks, just as in the equiprobable case; in addition, they are constructed to be orthogonal to the first component. Similar sets of scores were discussed by Barton[20] in the context of smooth tests for grouped data.

Specifically, let q_k be the overall proportion of observations that fall into category k. For example, for the post-etch window size data in Table 11.1, $(q_1, ..., q_5) = (.48, .09, .13, .19, .11)$. Let

$$\tau_k = \sum_{j=1}^{k-1} q_j + q_k/2 \tag{17}$$

so that τ_k is proportional to the midrank for category k. The first set of scores (location scores) is

$$l_k = \tilde{\tau}_k \Big/ [\sum_{j=1}^{K} q_j \tilde{\tau}_j^2]^{1/2}, \quad k = 1, ..., K \tag{18}$$

where

$$\tilde{\tau}_k = \tau_k - \sum_{j=1}^{K} q_j \tau_j = \tau_k - .5.$$

The second set of scores (dispersion scores) is

$$d_k = e_k \Big/ [\sum_{j=1}^{K} q_j e_j^2]^{1/2}, \quad k = 1, ..., K, \tag{19}$$

where

$$e_k = l_k(l_k - \sum_{j=1}^{K} q_j l_j^3) - 1.$$

We see from (18) and (19) that

$$\sum_{k=1}^{K} q_k \, l_k = \sum_{k=1}^{K} q_k \, d_k = \sum_{k=1}^{K} q_k l_k \, d_k = 0 \tag{20}$$

and

$$\sum_{k=1}^{K} q_k \, l_k^2 = \sum_{k=1}^{K} q_k \, d_k^2 = 1. \tag{21}$$

These scores can be used to analyze multifactor contingency table data as in Table 11.1. To do this, one multiplies the frequencies from each category by the location/dispersion scores and sums them across the categories to obtain a location/dispersion pseudo observation for the ith experimental run. These pseudo observations can then be analyzed using a standard ANOVA program. More specifically, consider a particular factor, say factor A, at I levels, and let $(Y_{i+k}, i=1,...,I, k=1,...,K)$ denote the counts in category k for the ith level of A summed across the levels of all the other factors. Then, the location pseudo observation for the ith level of factor A, summed across the levels of the other factors, is

$$L_i = \sum_{k=1}^{K} l_k Y_{i+k}.$$

Similarly, the dispersion pseudo observation is

$$D_i = \sum_{k=1}^{K} d_k Y_{i+k}.$$

The location sum of squares for factor A is then given by

$$SS_A(l) = \sum_{i=1}^{I} L_i^2 / n_i,$$

and the dispersion sum of squares is given by

$$SS_A(d) = \sum_{i=1}^{I} D_i^2 / n_i.$$

Here n_i denotes the total number of observations at the ith level of A. In multifactor situations, both $SS_A(l)$ and $SS_A(d)$, test for the effect of factor A by collapsing the data across the levels of the other factors.

Under the null hypothesis that none of the factors have any effects, the conditional distribution of the contingency table data, conditional on the column averages $\{q_k\}_{k=1}^K$, is given by a generalized multiple hypergeometric distribution (Lehmann[17]). It follows from the moments of this distribution and from (20) and (21) that, under the null hypothesis, the location and dispersion pseudo observations, L_i and D_i, have mean zero and are uncorrelated. Furthermore, the variance of the pseudo observation from the ith level is $n_i(N - n_i)/(N - 1)$ and the covariance of the observations from the ith and jth levels is $-n_i n_j/(N - 1)$. Here N denotes the total number of observations. The aforementioned properties also hold unconditionally. The non-zero correlations between the pseudo observations from different levels and the factors $(N - n_i)/(N - 1)$ in their variances are caused by the fact that they have to add to zero. Both L_i and D_i are analogous to the deviations $(Y_i - \bar{Y})$ in ANOVA. They can be analyzed by ignoring the covariances and taking the variance as $n_i/(1 - 1/N) \approx n_i$. This shows that, under the null hypothesis, the location and dispersion sum of squares $SS_A(l)$ and $SS_A(d)$ will be approximately distributed as χ_{I-1}^2 where I denotes the number of levels of the factor, and they will be approximately independent. It should be noted that these results hold only under the null hypothesis that none of the factors have any effects. A more careful analysis would require one to look at the distributions in the presence of other effects. This is, however, rather involved.

In Table 11.5, I compare the powers of these statistics with P and T using the same simulation set up described in Section 11.4. Note that the location scores l are asymptotically optimal for the logistic model (15) (Hajék and Sidak[21]). A comparison of the values in Table 11.5 for the configuration $(p_{.1}, ..., p_{.5}) = (.1, .15, .3, .2, .25)$ with the values in Table 11.4 shows that the powers of $SS(l)$ and $SS(d)$ are comparable with those of the first two components, Z_1^2 and Z_2^2, of T. The second configuration $(p_{.1}, ..., p_{.5}) = (.48, .09, .13, .19, .11)$ corresponds to the overall column proportions of the post-etch window size data in Table 11.1. In this case, almost half of the probability is concentrated in one column, and, as is to be expected, the powers of all the statistics are lower. $SS(l)$ and $SS(d)$ continue to have good power in detecting location and scale shifts respectively.

Table 11.6 gives the results from reanalyzing the post-etch window size data using these location and dispersion scores. In interpreting the mean squares in Table 11.6, I recommend (as does Taguchi) using the MS as evidence of the importance of an effect and not as a formal test of hypothesis. Following this philosophy, we see that the MS's for location give the same results as Taguchi's accumulation analysis in Table 11.2. This is to be expected, given the results in Sections 11.4 and 11.5. The accumulation analysis method, however, overlooks the important dispersion effects in Table 11.5: etch time (I), bake time (E) and aperture (F).

One other advantage of the scoring method in this context is that it allows one to easily identify the "optimal" level of a factor identified as being important. For example, factor I (etch time) is at three levels. To determine which level minimizes dispersion, I average the dispersion pseudo-observations at each level.

The level with the lowest value is the optimal one. This shows that for factor I the standard etching time is the proper setting; over-etching leads to increased variability in the window sizes.

TABLE 11.5. POWER COMPARISONS OF P, T, $SS(l)$, and $SS(d)$. The power computations are for $\gamma_A = .5$, $\gamma_B = .25$ and $\gamma_{AB} = 0$ in the location [scale] model (4.9) [(4.10)].

Statistics	$(p_{.1}, ..., p_{.5}) = (.1, .15, .3, .2, .25)$						$(p_{.1}, ..., p_{.5}) = (.48, .09, .13, .19, .11)$					
	Location Model			Scale Model			Location Model			Scale Model		
	γ_A	γ_B	γ_{AB}	γ_A	γ_B	γ_{AB}	γ_A	γ_B	γ_{AB}	γ_A	γ_B	γ_{AB}
P	.34	.14	.10	.68	.20	.11	.31	.13	.07	.44	.15	.10
T	.49	.19	.08	.59	.16	.09	.49	.20	.09	.32	.13	.10
$SS(l)$	.52	.21	.09	.12	.11	.11	.50	.20	.09	.17	.11	.10
$SS(d)$	.13	.11	.11	.81	.28	.07	.15	.11	.11	.58	.20	.09

TABLE 11.6. REANALYSIS OF POST-ETCH WINDOW SIZE DATA TO IDENTIFY LOCATION AND DISPERSION EFFECTS

Source	df	Mean Square for Location	Mean Square for Dispersion
A	1	7.50	1.30
BD	2	19.15	4.25
C	2	21.65	2.66
E	2	2.52	8.53
F	2	3.04	7.03
G	2	7.13	1.05
H	2	6.14	3.98
I	2	1.82	9.30

11.6 DISCUSSION

The idea behind choosing the scoring schemes is to judiciously decompose Pearson's chi-square statistic P into $K-1$ orthogonal components and select the two that contribute to most of the power under location and scale alternatives. Clearly, there are many ways of doing this (see Nair[4]) for components associated with cumulative chi-square tests other than Taguchi's accumulation analysis method, and for their relationships to Neyman-Barton smooth tests and optimal scores. An alternative to the scores proposed in Section 11.5 can be obtained by defining l_k

and d_k as in (18) and (19) with $\Phi^{-1}(\tau_k)$ instead of τ_k where $\Phi^{-1}(\cdot)$ is the inverse of the normal distribution function. These location scores give the grouped data version of the normal scores test (Lehmann[17]). The dispersion scores, in the equiprobable case, give the grouped data version of Klotz's[22] test. In most situations where we are interested in detecting only the really important factors, these scores would lead to the same conclusions as those in Section 11.5. The reanalysis of the post-etch window size data with these normal scores gave MS's whose magnitudes were very close to those in Table 11.6.

The main advantage of the scoring schemes considered in this article is their simplicity. It should be reiterated that there are a variety of other methods in the literature for modeling and analyzing ordered categorical data (see Agresti[6]). Among these, McCullagh's[23] regression models for ordinal data appear particularly useful for quality improvement experiments. This involves the fitting of a parametric model—say logistic model—and estimating the location and dispersion effects by maximum likelihood.

Finally, note that, to the extent possible, careful consideration should be given to the choice of factor levels and the definitions of categories in the design stage. Improperly chosen factor levels and categories can drastically reduce the information content in an experiment. For example, suppose that when a factor is at its lowest level, it reduces the mean so low that all of the observations fall into the first category. From these observations, one cannot make any inferences about the effects of the other factors. If the factor is at two levels, the efficiency of the experiment will be reduced by 50%. Further, if the categories are so coarse that all of the observations fall into one category, one cannot make any inferences separately about location and dispersion. As a general rule of thumb, it is best to choose the factor levels and define the categories (whenever possible) so that a roughly equal number of observations falls into each category.

11.7 SUMMARY

The goal in industrial experiments for quality improvement is to identify the location and dispersion effects associated with the design factors and determine the optimal factor levels to improve process quality. There are simpler and more efficient procedures than the accumulation analysis method for detecting location effects from ordered categorical data. Moreover, the method performs poorly in identifying the dispersion effects. Since it confounds the two effects, one cannot separate factors that affect location from those that affect variability. The scoring schemes suggested in this article can be used to separately identify the location and dispersion effects. They are intended as simple, approximate procedures. They can be supplemented with more sophisticated techniques, depending on the needs and statistical training of the user and the availability of software.

ACKNOWLEDGMENTS

The author is grateful to Peter McCullagh for valuable discussions and for assistance with the simulation program; to Madhav Phadke for introducing him to accumulation analysis; and to Mike Hamada, Jeff Wu, and two referees for useful comments. Thanks are also due to Jim Landwehr, Colin Mallows, and John Tukey for comments on an earlier version of the paper.

[Discussions of this article and the author's response appear in
Technometrics vol. 28, no. 4 (November 1986): 292-311.]

APPENDIX

We briefly outline an asymptotic analysis that shows why the two sets of scores associated with the first two components of T as well as those considered in Section 11.5 have good power against location and scale alternatives respectively.

Consider the $I \times K$ table in Section 11.3 with frequencies Y_{ik}, $i = 1, ..., I$, $k = 1, ..., K$, and $n = \sum_{k=1}^{K} Y_{ik}$. Let $p_k = F(x_k) - F(x_{k-1})$, $k = 1, ..., K$, with F supported on $(-\infty, \infty)$ and $x_0 = -\infty$, $x_K = +\infty$. Let $f(x)$ be the density and assume $\lim_{x \to \pm\infty} f(x) = \lim_{x \to \pm\infty} x f(x) = 0$. Suppose the cell probabilities are given by the following location-scale model

$$p_{ik} = F\left[\frac{x_k - \mu_i}{\sigma_i}\right] - F\left[\frac{x_{k-1} - \mu_i}{\sigma_i}\right] \tag{22}$$

for $i = 1, ..., I$ and $k = 1, ..., K$. Without loss of generality, assume $\sum_{i=1}^{I} \mu_i = 0$ and $\sum_{i=1}^{I} \beta_i = 0$ where $\beta_i = \log(\sigma_i)$.

Consider now a sequence of contiguous alternatives in (22) with $\mu_i^n = n^{-\frac{1}{2}} \mu_i$ and $\beta_i^n = n^{-\frac{1}{2}} \beta_i$. Then up to first order terms,

$$p_{ik}^n = p_k - n^{-\frac{1}{2}} [\mu_i \eta_k + \beta_i \zeta_k] \tag{23}$$

where $\eta_k = f(x_k) - f(x_{k-1})$ and $\zeta_k = x_k f(x_k) - x_{k-1} f(x_{k-1})$. Note that $\sum_{k=1}^{K} \eta_k = \sum_{k=1}^{K} \zeta_k = 0$.

Let $(l_1, ..., l_K)$ and $(d_1, ..., d_K)$ denote, respectively, the location and dispersion scores in (18) and (19) or the scores associated with the first two components of T. Let $L_i = n^{-\frac{1}{2}} \sum_{k=1}^{K} l_k Y_{ik}$ and $D_i = n^{-\frac{1}{2}} \sum_{k=1}^{K} d_k Y_{ik}$ denote, respectively, the standardized location and dispersion pseudo-observations. Then, it can be shown (Nair[4]) that, as $n \to \infty$, L_i and D_i converge in distribution to independent normal random variables with variance one and means

$$m_{1i} = -\mu_i \left[\sum_{k=1}^{K} l_k \eta_k\right] - \beta_i \left[\sum_{k=1}^{K} l_k \zeta_k\right] \tag{24}$$

and

$$m_{2i} = -\mu_i \left[\sum_{k=1}^{K} d_k \eta_k\right] - \beta_i \left[\sum_{k=1}^{K} d_k \zeta_k\right]. \tag{25}$$

Consider first the equiprobable case where $p_k = 1/K$ and $l_k = (k - \frac{1}{2} - K/2)$. We have $l_k = -l_{K-k+1}$ and $d_k = d_{K-k+1}$. Suppose $f(x)$ is symmetric. Then $\eta_k = -\eta_{K-k+1}$ and $\zeta_k = \zeta_{K-k+1}$. So, $\sum_{k=1}^{K} l_k \zeta_k = 0$ in (24) and m_{1i} depends only on the location parameter μ_i. Similarly, $\sum_{k=1}^{K} d_k \eta_k = 0$ in (25) so that m_{2i} depends only on the scale parameter β_i. Since we are modeling the location-scale family in (1) obtained after some transformation, it is not unreasonable to assume that $f(x)$ is approximately symmetric.

In the general nonequiprobable case, $\sum_{k=1}^{K} l_k \zeta_k$ in (24) will not equal zero so that the power associated with the location scores will depend to some extent on the scale parameter β_i. However, note that the location scores l_k's in (18) are linear in the mid-ranks while the dispersion scores d_k's in (19) are quadratic. So $\sum_{k=1}^{K} l_k \zeta_k$ in (24) will be small compared to $\sum_{k=1}^{K} l_k \eta_k$ so that m_{1i} depends primarily on μ_i. Similarly, $\sum_{k=1}^{K} d_k \eta_k$ in (25) will be small relative to $\sum_{k=1}^{K} d_k \zeta_k$ so that m_{2i} depends mostly on the scale parameter β_i.

REFERENCES

1. Taguchi, G. 1966. *Statistical Analysis*. Tokyo: Maruzen Publishing Company. (Japanese)

2. Taguchi, G. 1974. A New Statistical Analysis for Clinical Data, the Accumulating Analysis, In Contrast With the Chi-Square Test. *Saishin Igaku (The Newest Medicine)* **29**:806-813.

3. Phadke, M. S., R. N. Kackar, D. V. Speeney, and M. J. Grieco. 1983. Off-line Quality Control in Integrated Circuit Fabrication Using Experimental Design. *The Bell System Technical Journal* **62**:1273-1310.
   (Quality Control, Robust Design, and the Taguchi Method; Article 6)

4. Nair, V. N. 1986. Components of Cumulative Chi-Square Type Tests for Ordered Alternatives in Contingency Tables. *Statistical Research Reports #18.* Murray Hill, N.J.: AT&T Bell Laboratories.

5. Mood, A. M. 1954. On the Asymptotic Efficiency of Certain Nonparametric Two-Sample Tests. *Ann. Math. Stat.* **25**:514-22.

6. Agresti, A. 1984. *The Analysis of Ordinal Categorical Data*. New York: Wiley.

7. Box, G. E. P. 1984. Recent Research in Experimental Design for Quality Improvement with Applications to Logistics. *Technical Report #2774.* Madison, Wis.: MRC, University of Wisconsin-Madison.

8. Kackar, R. N. 1985. Off-line Quality Control, Parameter Design, and the Taguchi Method (with discussion). *Journal of Quality Technology* **17** (Oct.):175-246.
   (Quality Control, Robust Design, and the Taguchi Method; Article 4)

9. León, R. V., A. C. Shoemaker, and R. N. Kackar. 1985. Performance Measures Independent of Adjustment: An Alternative to Taguchi's Signal to Noise Ratio. *AT&T Bell Laboratories Technical Memorandum.*
   (Quality Control, Robust Design, and the Taguchi Method; Article 12)

10. Taguchi, G. 1976, 1977. *Experimental Design*. Vols. 1 and 2. Tokyo: Maruzen Publishing Company. (Japanese)

11. Taguchi, G., and Y. Wu. 1980. *Introduction to Off-line Quality Control*. Nagoya, Japan: Central Japan Quality Control Association.

12. Hartley, H. O., and K. S. E. Jayatillake. 1973. Estimation for Linear Models with Unequal Variances. *J. Amer. Stat. Assoc.* **68**:189-192.

13. Cochran, W. G. 1950. The Comparison of Percentages in Matched Samples. *Biometrika* **37**:256-66.

14. Pitman, E. J. G. 1937. Significance Tests Which May Be Applied to Samples from any Population. *Suppl. J. R. Statist. Soc.* **4**:119-130.

15. Light, R. J., and B. H. Margolin. 1971. Analysis of Variance for Categorical Data. *J. Amer. Stat. Assoc.* **66**:534-44.

16. Taguchi, K., and C. Hirotsu. 1982. The Cumulative Chi-Squares Method Against Ordered Alternatives in Two-Way Contingency Tables. *Rep. Stat. Appl. Res., JUSE* **29**:1-13.

17. Lehmann, E. L. 1975. *Nonparametrics: Statistical Methods Based on Ranks.* San Francisco: Holden-Day, Inc.

18. Sukhatme, B. V. 1958. Testing the Hypothesis that Two Populations Differ Only in Location. *Ann. Math. Stat.* **29**:60-78.

19. Bross, I. D. J. 1958. How to Use Ridit Analysis. *Biometrics* **14**:18-38.

20. Barton, D. E. 1955. A Form of Neyman's Ψ_k^2 Test of Goodness of Fit Applied to Grouped and Discrete Data. *Skand. Aktuartidskr.* **38**:1-16.

21. Hajék, J., and Z. Sidák. 1967. *Theory of Rank Tests.* New York: Academic Press.

22. Klotz, J. 1962. Nonparametric Tests for Scale. *Ann. Math. Stat.* **33**:498-512.

23. McCullagh, P. 1980. Regression Models for Ordinal Data (with discussion). *J. R. Statistical Society*, B, **42**:109-42. (1980)

12

PERFORMANCE MEASURES INDEPENDENT OF ADJUSTMENT: AN EXPLANATION AND EXTENSION OF TAGUCHI'S SIGNAL-TO-NOISE RATIOS

Ramon V. Leon,
Anne C. Shoemaker, and
Raghu N. Kackar

12.1 THE ROLE OF SIGNAL-TO-NOISE RATIOS IN PARAMETER DESIGN

12.1.1 PARAMETER DESIGN: REDUCING SENSITIVITY TO VARIATION

When the Ina Tile Company of Japan found that the uneven temperature profile of its kilns was causing unacceptable variation in tile size, it could have attempted to solve the problem with expensive modification of the kilns. Instead, it chose to make an inexpensive change in the settings of the tile design parameters to reduce sensitivity to temperature variation. Using a statistically planned experiment the company found that increasing the lime content of the clay from 1% to 5% reduced the tile size variation by a factor of 10 (see Taguchi and Wu[1]).

This simple example illustrates the method of parameter design for quality engineering. Parameter design is the operation of choosing settings for the design parameters of a product or manufacturing process to reduce sensitivity to noise. Noise is hard-to-control variability affecting performance. For example, all of the following are considered noise: deviations of raw materials from specifications, changes in the manufacturing or field operation environment such as temperature or humidity, drift of parameter settings over time, and deviation of design parameters from their nominal values because of manufacturing variability.

In parameter design, noise is assumed to be uncontrollable. After parameter design, if the loss caused by noise is still excessive, the engineer may proceed to control noise with relatively expensive countermeasures, such as the use of higher-grade raw materials or higher-precision manufacturing equipment. (For a more detailed discussion of parameter design and the subsequent design stage, tolerance design, see Taguchi and Wu[1].)

12.1.2 A FORMALIZATION OF A BASIC PARAMETER DESIGN PROBLEM

Figure 12.1 shows a block diagram representation of a simple parameter design problem like the tile example. For a given setting of design parameters, θ, noise N produces a characteristic output Y. The output is determined by the transfer function, $f(N; \theta)$. The noise is assumed to be random; hence, the output will be random. A loss is incurred if the output differs from a target t, which represents the ideal output.

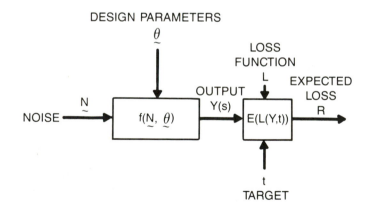

Figure 12.1. A Block Diagram Representation of a Simple Parameter Design Problem. The output Y is determined by the noise N through the transfer function f. The transfer function depends on the design parameters θ. Loss is incurred if the output is not equal to the target t.

The goal of parameter design is to choose the setting of the design parameter θ that minimizes average loss caused by deviations of the output from target.

This average loss is given by

$$R(\theta) = E_N L(Y,t)$$

where L is a loss function.

Taguchi (see Taguchi;[2] Taguchi and Phadke[3]) called this problem the *static* parameter design problem because the target is fixed. As we show later, other parameter design problems are generalizations of the static problem.

Taguchi recommended that loss be measured using the quadratic loss function

$$L(y,t) = K(y-t)^2 \tag{1}$$

where K is determined using a simple economic argument (see Taguchi and Wu[1]). Yet, as we will see in Section 12.1.3, in practice he optimized different measures called signal-to-noise (SN) ratios.

12.1.3 TAGUCHI'S APPROACH TO PARAMETER DESIGN

To solve the problem just described, Taguchi[2] generally divided the design parameters into two groups, $\theta = (d,a)$, and used the following two-step procedure (see Phadke[4]):

Procedure 1.
Step 1. Find the setting $d = d^*$ that maximizes the SN ratio.
Step 2. Adjust a to a^* while d is fixed at d^*. $\tag{2}$

The division of design parameters into two groups is motivated by the idea that the parameters a are fine-tuning adjustments that can be optimized after the main design parameters d are fixed. Taguchi (see Phadke[4]) claimed that this two-step approach has the advantage that, once a design (d^*,a^*) is established, certain changes to product or process requirements can be accommodated by changing only the setting of the adjustment parameter, a. The initial setting of $d = d^*$ remains optimal. For instance, in the Ina Tile example mentioned earlier, the clay formulation could be chosen to minimize an SN ratio measurement of tile size variation, and then the tile mold size could be used to adjust the average tile size to target.

Recognizing that the appropriate performance measure depends on the characteristics of the problem, Taguchi (see Taguchi and Phadke[3]) classified parameter design problems into categories and defined a different SN ratio for each category. For example, one of these categories consists of all static parameter design problems.

In general, however, Taguchi[2] gave no justification for the use of the SN ratios and no explanation of why the two-step procedure that he recommended will minimize average loss. In this article, we explore this point, investigating the connection between a two-step procedure similar to Procedure 1 and parameter design problems similar to the one stated in Section 12.1.2.

12.1.4 OVERVIEW

In Section 12.2, we examine performance measures in the most common parameter design situation, the static problem of Section 12.1.2. In Section 12.2.1 we show that for this problem Taguchi's SN ratio and corresponding two step procedure is valid assuming quadratic loss and a particular multiplicative transfer function model. We also discuss the danger of using this SN ratio and corresponding two step procedure when a different type of transfer function model may hold. Taguchi's recommendation to use a single SN ratio for all static parameter design problems, however, may be motivated by a belief that the transfer-function model for most engineering systems is multiplicative in form. In Section 12.2.2 we see how a multiplicative model arises from a general situation in which noise is manufacturing variation, deterioration, or drift in design parameter values.

In Section 12.3 we discuss block diagram representations of more general parameter design problems. These generalizations include "dynamic" problems, in which there is a "signal" input variable that can be used to dynamically control the output. In that section we also generalize the recipe followed in Section 12.2 to derive a two-step procedure equivalent to Procedure 1. Since the appropriate performance measure to use in Step 1 of this procedure is not always equivalent to an SN ratio, we suggest the term *performance measure independent of adjustment* (PerMIA).

Sections 12.4 and 12.5 are devoted to generic examples of dynamic parameter design problems, for which we follow the recipe to obtain performance measures independent of adjustment and corresponding two-step procedures. These examples show that Taguchi's various SN ratios for dynamic problems arise from certain types of transfer-function models under quadratic loss. If different types of transfer functions hold, however, a convenient two-step procedure may still be obtained if a performance measure different from the SN ratio is used. In fact, use of the SN ratio in these situations is incorrect and may not lead to good design parameter settings. Section 12.4 focuses on dynamic problems in which both input and output variables are continuous. Section 12.5 focuses on a binary-input-binary-output dynamic problem. Finally, in Section 12.6, we discuss how adjustment parameters may arise in practice and the benefits of the two-step procedure made possible by their existence.

12.2 PERFORMANCE MEASURES FOR A STATIC PARAMETER DESIGN PROBLEM

12.2.1 EXPLANATION AND EXTENSION OF TAGUCHI'S SN RATIO

Taguchi's SN ratio for the static parameter design problem of Figure 12.1 is

$$SN = 10 \log_{10} \frac{E^2 Y}{Var\ Y} \ , \tag{3}$$

where EY and $Var\ Y$ are, respectively, the mean and variance of Y (see Taguchi and Phadke[3]).

We show that if the transfer function in Figure 12.1 is given by

$$Y = \mu(\mathbf{d},\ \mathbf{a})\varepsilon(\mathbf{N},\ \mathbf{d}) \tag{4}$$

where $EY = \mu(\mathbf{d},\ \mathbf{a})$ is a strictly monotone function of each component of $\mathbf{a}$ for each $\mathbf{d}$, then the use of SN ratio (3) in a two-step procedure leads to minimization of quadratic loss. [Note that $EY = \mu(\mathbf{d},\ \mathbf{a})$ implies that $E_\varepsilon(\mathbf{N},\ \mathbf{d}) = 1$.] Model (4) could hold, for example, if the noise affects the output, Y, uniformly over increments of time or distance. For example, if a film is being deposited on a quartz plate, the thickness of the film will tend to vary a certain amount for each micron of film surface deposited. That is, a 10 micron-thick film will have 10 times the standard deviation of a 1-micron-thick film. For another example, see Section 12.2.2.

Note that model (4) essentially says that $\dfrac{Var\ Y}{E^2 Y}$ does not depend on $\mathbf{a}$ (or approximately, for log Y, that $\mathbf{a}$ affects location but not dispersion).

The argument linking the SN ratio (3) to quadratic loss is straightforward. We first note that we can find $(\mathbf{d}^*, \mathbf{a}^*)$ such that

$$R(\mathbf{d}^*, \mathbf{a}^*) = \min_{\mathbf{d}\mathbf{a}} R(\mathbf{d},\ \mathbf{a})$$

by following a general two-stage optimization procedure (Procedure 2). Before stating Procedure 2, we will define $P(\mathbf{d})$ by the equation $P(\mathbf{d}) = \min_{\mathbf{a}} R(\mathbf{d},\ \mathbf{a})$.

Procedure 2.
Step 1. Find $\mathbf{d}^*$ that minimizes

$$P(\mathbf{d}) = \min_{a} R(\mathbf{d},\mathbf{a}) \ . \tag{5}$$

Step 2. Find $\mathbf{a}^*$ that minimizes $R(\mathbf{d}^*, \mathbf{a})$.

241

To see that this procedure always works [provided that (a^*,d^*) exists] let (d,a) be any arbitrary parameter values and (d^*,a^*) be the result of Procedure 2. Then note that

$$R(d,a) \geq \min_a R(d,a)$$

$$\geq \min_a R(d^*,a) \text{ (by stage 1)}$$

$$= R(d^*,a^*) \text{ (by stage 2)} .$$

Even though this procedure is always possible, it usually is not useful. If there is a shortcut for calculating $P(d) = \min_a R(d,a)$, however, this two-step procedure may be the preferred minimization method.

If we are using quadratic loss and model (4) holds, then $P(d)$ is a decreasing function of Taguchi's SN ratio (3). This means that in this case Taguchi's two-step procedure, described in Section 12.1.3, is equivalent to Procedure 2.

To see this, note that under model (6) and quadratic loss,

$$R(d,a) = \mu^2(d,a)\sigma^2(d) + (\mu(d,a) - t)^2 \tag{6}$$

where $\sigma^2(d) = Var\ E(N,d)$. [For convenience choose $K=1$ in loss function (1).] Setting

$$\frac{\partial R(d,a)}{\partial a} = \frac{2\partial\mu(d,a)}{\partial a} \left\{ \mu(d,a)[1 + \sigma^2(d)] - t \right\}$$

equal to 0, we get

$$\mu(d,a^*(d)) = \frac{t}{1+\sigma^2(d)} ,$$

where $a^*(d)$ is defined by $R(d,a^*(d)) = \min_a R(d,a)$. Then, substituting into formula (6), we get

$$P(d) = \frac{t^2\sigma^2(d)}{1+\sigma^2(d)} .$$

But

$$SN = 10 \, log_{10} \frac{E^2 Y}{Var \, Y}$$

$$= 10 \, log_{10} \frac{\mu^2(\mathbf{d},a)}{\mu^2(\mathbf{d},a)\sigma^2(\mathbf{d})}$$

$$= -10 \, log_{10} \, \sigma^2(\mathbf{d}) \, ,$$

and $\dfrac{t^2\sigma^2(\mathbf{d})}{1+\sigma^2(\mathbf{d})}$ is an increasing function of $\sigma^2(\mathbf{d})$.

Hence the two-step procedure (Procedure 2) for solving the parameter design problem is equivalent to the following:

Procedure 3.

Step 1. Find $\mathbf{d}^*$ that maximizes the *SN* ratio

$$SN = 10 \, log_{10} \frac{E^2 Y}{Var \, Y}$$

Step 2. Find $\mathbf{a}^*$ such that

$$\mu(\mathbf{d}^*, \, \mathbf{a}^*) = \frac{t}{1+\sigma^2(\mathbf{d}^*)} \, .$$

With transfer function (4), the use of the *SN* ratio (3) leads to minimization of quadratic loss. (Note that even if the minimization of the expected squared error loss is constrained by the requirement that the mean be equal to t, Step 1 remains the same. What changes is Step 2, which now becomes "adjust to t.")

Although use of the *SN* ratio (3) under transfer function (4) is justified by the preceding argument, there may be other measures that are better in practice. In particular, if information about the loss function is not precise, there may be no reason to prefer quadratic loss over, say, quadratic loss on the log scale:

$$L(y,t) = (log \, y - log \, t)^2 \, .$$

Under this loss function, transfer function (4) leads to use of $Var(log \, Y)$ in Step 1 of Procedure 2. To see this, note that

$$R(\mathbf{d},a) = E(log \, Y - log \, t)^2$$

$$= Var(log \, Y) + [E(log \, Y) - log \, t]^2 \, .$$

Since, from model (4)

$$Var(\log Y) = Var(\log \varepsilon \ (\mathbf{N},\mathbf{d}))$$

is a function of $\mathbf{d}$ alone, it follows that

$$\min_{\mathbf{a}} R(\mathbf{d},\mathbf{a}) = Var(\log Y) .$$

The quantity $\log Var(\log Y)$ was suggested by Box[5] as an alternative to the *SN* ratio of Taguchi. When there is no practical reason to prefer either of the preceding loss functions, Box's suggestion should probably be followed, because the estimate of $\log Var(\log Y)$ usually has better statistical properties than those of the *SN* ratio (see Box[5]). (Note, however, that problems may arise with this loss function if t is close to 0.)

We also note that if the transfer function in (4) is replaced by

$$Y = \mu(\mathbf{d},\mathbf{a}) + \varepsilon(\mathbf{N},\mathbf{d}) \tag{7}$$

with $E_\varepsilon(\mathbf{N},\mathbf{d}) = 0$ and $L(y,t) = (y-t)^2$, then $Var\ Y$ should be used instead of the *SN* ratio (3) in the two-step procedure. Note that the SN ratio is not independent of the adjustment parameter, $\mathbf{a}$, under model (7). This example illustrates that blanket use of the SN ratio in static problems, as Taguchi and Phadke seem to advocate, could lead to far from optimal design parameter settings.

12.2.2 EXAMPLE: STATIC PROBLEM WITH VARIATION IN DESIGN—PARAMETER VALUES

Suppose that the output, Y, of the system is given by

$$Y = A^p g(\mathbf{D}) \quad (-\infty<p<\infty) \tag{8}$$

where $(A, \mathbf{D})$ are random design parameters settings that, because of manufacturing variability or deterioration, are random variables with means equal to the nominal settings (a, d) set by the designer. (Taguchi[2] referred to this kind of noise as "inner" noise.) Model (8) with $p=1$ (suggested in this case by Paul Sherry, a quality engineer at AT&T Bell Laboratories) frequently occurs in applications. In these applications parameter a is called a *scale* parameter because it can be used to change the "scale" of the product or process. Examples of scale parameters [and model (8)] are mold size in tile fabrication, deposition time in surface film deposition, mask dimension in integrated circuit fabrication, and exposure time in window photolithography.

Engineers frequently observe that variability of the actual scale parameter value (as measured by the standard deviation) is proportional to nominal settings. Hence we assume that

$$Var\ A = k^2 a^2 \qquad k \geq o\ .$$

To see that model (8) is a special case of model (4), represent A by

$$A = aZ\ ,$$

where Z is a random variable with $EZ=1$ and $Var\ Z = k^2$. Then

$$Y = a^p Z^p g(\mathbf{D})$$

$$= \mu(\mathbf{d},a)\varepsilon(\mathbf{N},\mathbf{d})\ ,$$

where $\mu(\mathbf{d},a) = a^p E[Z^p g(\mathbf{D})]$ and $\varepsilon(\mathbf{N},\mathbf{d}) = \dfrac{Z^p g(\mathbf{D})}{E[Z^p g(\mathbf{D})]}$. From the result of Section 12.2.1, we see that under model (8) Taguchi's two stage procedure leads to the minimization of average quadratic loss.

12.3 GENERAL MODEL FOR PARAMETER DESIGN PROBLEMS AND A CONVENIENT TWO-STEP SOLUTION

As mentioned in Section 12.1.2, a convenient way of thinking about parameter design problems is in terms of block diagrams similar to that in Figure 12.1. Another example of a block diagram representation is given in Figure 12.2.

In this case, for a given setting of the design parameters $\mathbf{d}$ and $\mathbf{a}$, the output Y is determined by an input signal s and by noise $\mathbf{N}$. Loss is incurred if the output is different from a target that may depend on the input signal. As before, the goal of parameter design is to choose the setting of the design parameters that minimizes expected loss,

$$R(\mathbf{d},a) = E_s E_N(L(Y,t(s)) \mid \mathbf{d},a)\ , \tag{9}$$

where the distribution of the signal, s, reflects the relative frequency of its different values.

This kind of system is called *dynamic* because the output and the target depend on a signal that is not fixed by the designer. For example, a measuring instrument such as a bathroom scale is a dynamic system because its output, a

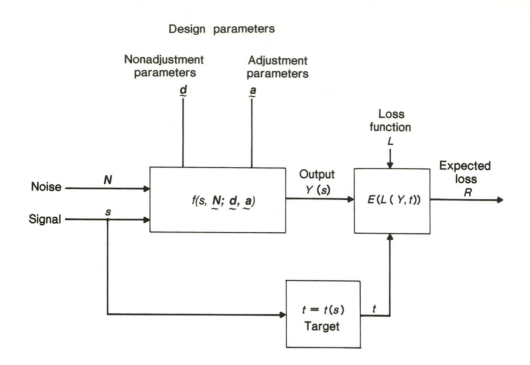

Figure 12.2. A Generic Parameter Design Problem. The output, Y(s), is determined by an input signal, s, and noise N, through a transfer function, f. The transfer function depends on the design parameters d and a. Loss is incurred if the output is not equal to the target value, $t(s)$.

weight reading, depends on the input signal, the actual weight of the person standing on the scale. The weight reading is also affected by noise factors such as temperature.

The block diagram in Figure 12.2 does not include every parameter design problem. For example, there may not be a signal, as we saw in the static problem of Section 12.1.2. Moreover, in the control problem of Section 12.4.2 the signal is determined by the target rather than vice versa. (The block diagram corresponding to this problem is shown in Figure 12.3.) The block diagram can be appropriately modified to represent most parameter-design problems, however. For example, Khosrow Dehnad of AT&T Bell Laboratories has suggested that for some problems, such as the design of an amplifier, the function $t(s)$ might be replaced by the class of all linear functions, since to avoid distortion an amplifier's output must be close to a linear function of its input.

The objective of parameter design is to find the values of design parameters (d,a) that minimize average loss $R(d,a)$. This can conveniently be done in many parameter design problems by following the two-stage Procedure 2. As mentioned in Section 12.2, this two-step procedure is always possible though it may not be useful. Under certain models and loss functions, however, $P(\mathbf{d})$ is equivalent to Taguchi's SN ratio and Procedure 2 is equivalent to Procedure 1. In Section 12.2 we demonstrated this claim for the static parameter design problem. In the rest of the article we show other examples in which $P(\mathbf{d})$ is equivalent to the SN ratio given by Taguchi.

Under other models and other loss functions, however, the same two-step procedure is possible, but $P(\mathbf{d})$ is not equivalent to Taguchi's SN ratio. For this reason, when a procedure like Procedure 2 can be conveniently used, we call $P(\mathbf{d})$ (or a monotone function of it) a PerMIA, rather than a SN ratio.

For fixed $\mathbf{d}$ we let $\mathbf{a}^*(\mathbf{d})$ be the value of $\mathbf{a}$, where $R(\mathbf{d},\mathbf{a})$ is minimized; that is,

$$P(\mathbf{d}) = \min_{\mathbf{a}} R(\mathbf{d},\mathbf{a}) = R(\mathbf{d},\mathbf{a}^*(\mathbf{d})). \tag{10}$$

12.4 PERFORMANCE MEASURES FOR DYNAMIC PARAMETER DESIGN PROBLEMS

As discussed in Section 12.3, parameter design problems in which the output and the target depend on a signal are called dynamic. As in all parameter design problems, the objective is to find the settings of design parameters of a product or process so that loss caused by deviation of output from target is minimized. In this section, we consider two examples of dynamic parameter design problems. The first example concerns the design of a measuring instrument, such as a spring scale. The performance measure is independent of calibration parameters, allowing comparison of scale designs without calibration. In the second example, we derive a performance measure for a generic control problem, in which the objective is to design a product so that the output can be varied over a certain interval by changing the input signal. In Appendix A, we discuss in detail a specific control problem involving the design of a truck-steering mechanism.

Both of these examples are of the type that Taguchi and Phadke[3] call continuous-continuous dynamic classified parameter design problems because the input and output are continuous variables. Taguchi[2] suggested a single SN ratio for all such problems. To see what this SN ratio is, let the general form of the continuous-continuous dynamic problem be given by the formula

$$Y = \alpha + \beta's + \varepsilon', \tag{11}$$

where s is a continuous signal controlling a continuous output Y, and ε' is some deviation from linearity. Then Taguchi (see Taguchi and Phadke[3]) gave the following SN ratio:

$$SN = \log_{10} \frac{\beta'^2}{VAR(\varepsilon')}. \tag{12}$$

247

This SN ratio is defined by formulas (11) and (12) regardless of the actual form of the model for Y.

In the first example considered in this section we follow Procedure 2 for finding a PerMIA. We show that the SN ratio (12) is a PerMIA under a certain special model, and therefore its use leads to minimization of squared-error loss. If a different model holds, however, use of this SN ratio is not appropriate. In the second example, the SN ratio is not independent of adjustment parameters, but a heuristic argument is presented that suggests that the SN ratio may sometimes lead to the optimum.

12.4.1 AN EXPLANATION OF AN SN RATIO: MEASURING INSTRUMENT EXAMPLE

Suppose a manufacturer wants to design a measuring instrument and that the dial reading Y of the instrument satisfies

$$Y = \alpha(\mathbf{d}, a_1) + \beta(\mathbf{d}, a_2)\ (\gamma(\mathbf{d})s + \varepsilon(\mathbf{N}; \mathbf{d})) \tag{13}$$

where

$$VAR\left[\varepsilon(\mathbf{N}; \mathbf{d})\right] = \sigma^2(\mathbf{d})$$

and

$$E\left[\varepsilon(\mathbf{N}; \mathbf{d})\right] = 0\ .$$

Here s is the true value of the measured quantity and Y is the reading on the instrument's dial.

To see why this model might be appropriate for a measuring instrument which can be calibrated, consider the case of a spring scale such as a postal scale. The scale consists of two parts, the sensor, which is the spring; and a dial, which translates the spring's compression into a weight reading. A model for the sensor part of the scale is

$$Z = \gamma(\mathbf{d})s + \varepsilon(\mathbf{N}; \mathbf{d}),$$

where Z is the compression (in inches) of the spring, s is the true weight of an object, $\mathbf{d}$ is the design parameters of the spring such as spring size and alloy and $\mathbf{N}$ is noises, such as imperfections in the spring, which affect compression. Since the location and spacing of markings on the dial can be chosen by the designer, a model for the dial reading, Y, is

$$Y = \alpha(\mathbf{d}, a_1) + \beta(\mathbf{d}, a_2)Z,$$

248

where a_1 and a_2 are adjustments chosen by the designer. Substituting for Z gives model (13).

For the general measuring instrument described by model (13), the desired value for the dial reading Y is equal to the true value s. Suppose loss is measured by $(Y - s)^2$; then the objective of parameter design is to find the setting of $(\mathbf{d},\mathbf{a})$ that minimizes

$$E_s \, E_N \left[(Y-s)^2 \mid \mathbf{d}, \mathbf{a} \right] \, .$$

In addition, suppose the designer requires that

$$(\mathbf{d},\mathbf{a}) \; \varepsilon \; \left\{ (\mathbf{d},\mathbf{a}): \; E_N(Y \mid \mathbf{d},\mathbf{a},s) = s \right\},$$

that is, the design must give an unbiased estimate of s. Then, to find the optimal setting, $(\mathbf{d}^*,\mathbf{a}^*)$, note that

$$\min_{\mathbf{d}} \min_{\mathbf{a}} E_s \, E_N \left[(Y-s)^2 \mid \mathbf{d}, \mathbf{a} \right]$$

$$= \min_{\mathbf{d}} \min_{\mathbf{a}} \; \left\{ \beta^2(\mathbf{d},a_2) Var_N(\varepsilon(\mathbf{N};\mathbf{d})) \right.$$

$$+ E_s \left[\alpha(\mathbf{d},a_1) + \beta(\mathbf{d},a_2) \; \gamma(\mathbf{d})s - s \right]^2 \Big\}$$

$$= \min_{\mathbf{d}} \; \frac{Var_N(\varepsilon(\mathbf{N};\mathbf{d}))}{\gamma^2(\mathbf{d})} \, .$$

The last of the preceding equalities holds because $a_1^*(\mathbf{d})$ and $a_2^*(\mathbf{d})$ must satisfy

$$\alpha(\mathbf{d},a_1^*(\mathbf{d})) = 0$$

and

$$\beta(\mathbf{d},a_2^*(\mathbf{d}))\gamma(\mathbf{d}) = 1$$

for each $\mathbf{d}$, under the unbiassedness constraint.

The function

$$\frac{Var_N(\varepsilon(N;d))}{\gamma^2(d)}$$

is a PerMIA because it does not depend on the adjustment parameters, a_1 and a_2. This PerMIA measures the performance of the measuring instrument as it would be after proper calibration.

Hence, parameter design can be done using the following two-step procedure:

Procedure 4.
Step 1. Find d^* that minimizes the PerMIA,

$$\frac{Var_N(\varepsilon(N;d))}{\gamma^2(d)} .$$

Step 2. Find a_1^* and a_2^* for which

$$\alpha(d^*, a_1^*) = 0$$

$$\beta(d^*, a_2^*) = \frac{1}{\gamma(d^*)} .$$

As seen by comparing formulas (11) and (13), $\beta' = \beta(d, a_2)$ and $\varepsilon' = \beta(d, a_2)\varepsilon$. So optimizing the PerMIA is equivalent to optimizing the SN ratio given in formula (12), the one suggested by Taguchi for all continuous-continuous parameter design problems. In fact, the derivation of the PerMIA given in this section is motivated by the discussion in Chapter 22 of Taguchi.[2]

Let us see what happens, however, if we retain the same loss function but change the model for the measuring instrument as follows:

$$Y = \alpha(d, a_1) + \beta(d, a_2) s + \varepsilon(N; d)$$

where

$$E(\varepsilon(N; d)) = 0$$

and

$$VAR(\varepsilon(N; d)) = \sigma^2(d) .$$

Then, retracing the previous argument, the PerMIA for the measuring-instrument parameter design can be shown to be $\sigma^2(\mathbf{d})$. But under this model the PerMIA is not equivalent to Taguchi's SN ratio. In fact, under this model the SN ratio depends on the adjustment parameter, a_2. Hence, using the SN ratio as if it were independent of both adjustment parameters could lead to unfair comparisons of alternative settings of the nonadjustment design parameters, $\mathbf{d}$.

We emphasize throughout this article that models and loss functions should be carefully stated for each parameter design problem, because the appropriate performance measure may be very dependent on them. The SN ratio formula (12) cannot be safely used for all continuous-continuous parameter design problems.

12.4.2 PERFORMANCE MEASURE FOR A CONTROL SYSTEM EXAMPLE

As an illustration of a control system, consider a truck's steering mechanism. The truck driver and the truck together make up a control system. The driver chooses a certain steering angle, $s(t)$, to accomplish a turn of radius t. The design of the steering mechanism must allow the driver to find a steering angle for any turning radius over some range. The possible steering angles may also be restricted to some interval. In addition, the turning radius of the truck resulting from a given steering angle should be consistent and not affected by noise conditions such as road surface, load distribution, and tire air pressure.

The parameter design of a truck's steering mechanism is considered in detail in Appendix A. In this section we present a generic control problem for which we postulate a model and derive a PerMIA. Although the model assumed here is different from the model that is derived for the steering mechanism in Appendix A, the PerMIAs turn out to be the same in both cases.

The block diagram in Figure 12.3 describes a general control problem. The input signal s is determined by the intended target t and the control strategy $s(t)$. For simplicity we assume that $s(t)$ is the unbiased control strategy given by

$$s(t) = \{s: E(Y(s) \mid \mathbf{d},\mathbf{a},s) = t\},$$

where $Y(s)$ is the system output corresponding to signal s. The objective of parameter design is to find the setting of the design parameters, $(\mathbf{d},\mathbf{a})$, which minimizes expected loss $R(\mathbf{d},\mathbf{a})$. A more general objective of the parameter design problem would be to minimize expected loss not only over the settings of the design parameters, but also over some set of permissible control strategies. In this article, however, we restrict that set to consist of only one strategy.

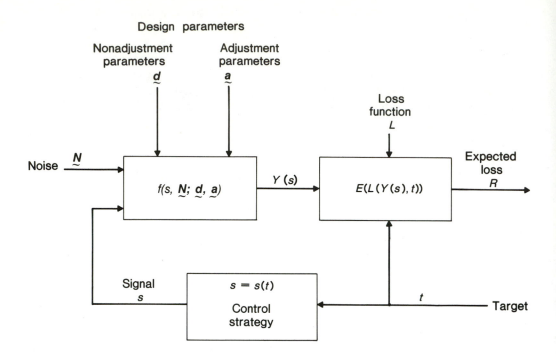

Figure 12.3. Parameter Design of a Control Problem. The input signal is determined by the control strategy $s(t)$ and the intended target, t. The object of the parameter design is to find the setting of the design parameters, (d,a), which minimizes expected loss $R(d,a)$. A more general objective of the parameter design problem would be to find the minimum expected loss not only over the design parameter settings, but also over the permissible control strategies.

For a given setting of the design parameters, (d,a), suppose the following model describes the relationship between the output Y, the signal s, and noise N:

$$Y = \alpha(d, a) + \beta(d)s + \varepsilon(N, d), \tag{14}$$

where

$$E(\varepsilon(N, d)) = 0$$

and

$$Var(\varepsilon(N, d)) = \sigma^2(d) .$$

Assuming that loss caused by deviation of Y from t is measured by $(Y - t)^2$, the objective is to find the setting of (d, a) which minimizes

$$E_t E_N \left[(Y - t)^2 \mid d, a \right] .$$

For any target t in (t_L, t_H) the unbiased control strategy dictates that $s = s(t)$ will be chosen from (s_L, s_H) so that $E(Y \mid d, a, s) = t$. This will be possible only if the slope $\beta(d)$ is such that

$$| \beta(d) | \geq \left| \frac{t_H - t_L}{s_H - s_L} \right| .$$

Consequently, the design parameters d must satisfy this constraint. Notice that there is no similar constraint for the y-intercept $\alpha(d, a)$ because, if the slope is steep enough, the adjustment variable, a, can be used to shift $E(Y \mid d, a, s)$ up or down.

It follows that the two-step decomposition of the parameter design is the following:

Procedure 5.
Step 1. Find d^* to minimize

$$\sigma^2(d)$$

subject to the constraint

$$\beta^2(d) \geq \left[\frac{t_H - t_L}{s_H - s_L} \right]^2 .$$

Step 2. Choose a^* so that

$$\alpha(d^*, a^*) + \beta(d^*)s,$$

$s \, \epsilon [s_L, s_H]$ can cover the target interval $[t_L, t_H]$.

The constraint on d is frequently important in practice, since values of d that tend to make the effects of the noise small,—that is, $\sigma^2(d)$ small—may also tend to make the effect of the signal small, that is, β small. As an extreme example, the truck's steering mechanism is least sensitive to noise if it is welded solid, but then the ability to steer is lost completely.

As mentioned previously, Taguchi and Phadke[3]) and Taguchi[2] recommend a different performance measure for all continuous-continuous dynamic problems, namely

$$S/N = 10 \log_{10} \frac{\beta^2(\mathbf{d})}{\sigma^2(\mathbf{d})} . \tag{15}$$

In the parameter design of a control system such as the design of a truck steering mechanism, however, the principal objective is to minimize the effect of noise on output while maintaining at least a *minimum* sensitivity to the signal. Maximizing SN in Equation (15) may approximately accomplish this goal, because values of $\mathbf{d}$ which make $\frac{\beta^2(\mathbf{d})}{\sigma^2(\mathbf{d})}$ large will tend to make $\sigma^2(\mathbf{d})$ small and make $\beta^2(\mathbf{d})$ large (and hence satisfy our constraint). It seems unnecessary, however, to try to make β much larger than the minimum, especially if that results in a larger value for σ^2.

12.5 PERFORMANCE MEASURE FOR A BINARY-INPUT-BINARY-OUTPUT DYNAMIC PROBLEM

A type of dynamic parameter-design problem which arises in the communications and electronics industries is the binary-input-binary-output problem. This case is identical to the continuous dynamic problem discussed in Section 12.4, except that both the input and output are binary rather than continuous.

In this section we will derive a performance measure for a simple binary-input-binary-output problem, the design of a binary transmission channel. We show that certain model assumptions lead to a performance measure independent of a channel adjustment parameter. This derivation reveals conditions under which the SN ratio that Taguchi proposed for binary-input-binary-output problems is independent of the adjustment parameter and leads to minimization of average loss.

12.5.1 PERFORMANCE MEASURE FOR A BINARY TRANSMISSION CHANNEL

The purpose of a binary channel is to transmit a 0-1 signal over a wire. The signal must actually be sent over the wire as a voltage, however, so a 0 is converted to a voltage μ_0 and a 1 to a voltage μ_1 ($\mu_0 \le \mu_1$). The voltages actually received are $\mu_0 + \sigma\varepsilon$ and $\mu_1 + \sigma\varepsilon$, because electrical disturbance in the wire adds a random noise ε to the signal. This noise is assumed to be normally distributed with mean 0 and variance 1 (see Chapter 8 of Pierce,[6] or Chapter 8 of Pierce and Posner[7]). The value of σ is a positive number that reflects the channel's reaction to the noise. The values of μ_0, μ_1, and σ depend on the settings of some design parameters, **d**. If the received voltage is less than some voltage threshold a, it is assumed that a 0 was sent. If the received voltage is greater than a it is assumed that a 1 was sent. Note that a is also a design parameter (see Figure 12.4).

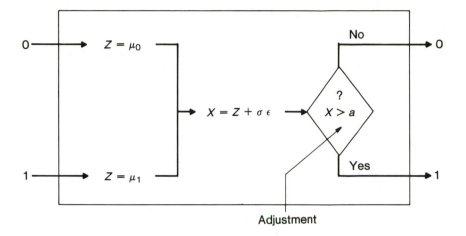

Figure 12.4. Schematic of Binary Channel. The engineer can send a known stream of 0's and 1's, and observe the resulting output of 0's and 1's. The engineer can change the misclassification probabilities by adjusting a.

To describe the problem in the framework of Figure 12.2, note that the signal s is the 0-1 input and the output Y is the received 0-1 message. Since in this example the purpose is to reproduce the signal, the target t is related to the signal by the identify function $t(s) = s$. The noise, **N**, is the electrical disturbance in the wire that affects the system through the function $\varepsilon = \varepsilon(\mathbf{N})$. The transfer function is as given in Figure 12.4, or in symbols,

$$Y = f(s, \mathbf{N}; \mathbf{d}, a)$$

$$= \begin{cases} 1 \text{ if } \mu_s + \sigma\varepsilon > a \\ 0 \text{ if } \mu_s + \sigma\varepsilon \leq a \end{cases}$$

where μ_0, μ_1 and σ are functions of the design parameters $\mathbf{d}$. The letter a stands, as before, for the voltage threshold. The loss function is given by

$$L(Y,t) = (Y-t)^2 = \begin{cases} 0 \text{ if } Y = t \\ 1 \text{ if } Y \neq t . \end{cases}$$

Assuming that 0s and 1s are sent with equal frequency the expected loss is given by

$$R(\mathbf{d},a) = E_s E_N(L(y,t) \mid \mathbf{d},a)$$

$$= P_r(s=0,Y=1) + P_r(s=1,Y=0)$$

$$= P_r(Y=1 \mid s=0)P_r(s=0) + P_r(Y=0 \mid s=1)P_r(s=1)$$

$$= \tfrac{1}{2}\,(p_0 + p_1)\,,$$

where $p_0 = p_0(\mathbf{d},a)$ and $p_1 = p_1(\mathbf{d},a)$ are, respectively, the probabilities of misclassifying a 0 and a 1.

The designers would like to compare several candidate settings of the design parameters, $(\mathbf{d},a)$, to find the one that minimizes average loss caused by misclassifying the signal. (More correctly, the engineer would like to maximize channel capacity. This criterion from information theory leads to a similar analysis. See Chapter 8 of Pierce.[6]) Assume that the engineers cannot observe voltages but that they can send a known sequence of 0s and 1s into the channel and observe the resulting output.

What is a convenient way to find the optimal value of the design parameters $(\mathbf{d},a)$? They can follow Procedure 2. To see how, first note that the error induced by noise is assumed to have a normal distribution, so

$$p_0 = \Phi(\frac{\mu_0-a}{\sigma})$$

and

$$p_1 = \Phi(\frac{a-\mu_1}{\sigma})$$

where a is the current voltage threshold and Φ is the standard normal cumulative distribution function. Then note that

$$R(\mathbf{d},a) = \frac{1}{2}\{\Phi(\frac{\mu_0-a}{\sigma}) + \Phi(\frac{a-\mu_1}{\sigma})\}$$

is minimized at

$$a^*(\mathbf{d}) = \frac{\mu_0(\mathbf{d}) + \mu_1(\mathbf{d})}{2}.$$

That is, for a given value of $\mathbf{d}$ the optimal value of the voltage threshold is half way between the two average transmission voltages (this is easy to prove).
    Using the fact that

$$\Phi(\frac{\mu_0-a^*(\mathbf{d})}{\sigma}) = \Phi(\frac{a^*(\mathbf{d})-\mu_1}{\sigma})$$

we then see that

$$R(\mathbf{d},a^*(\mathbf{d})) = \Phi\left[\frac{\dfrac{\mu_0+\mu_1}{2} - \mu_1}{\sigma}\right]$$

$$= \Phi\left[\frac{\mu_0-\mu_1}{2\sigma}\right]$$

$$= \Phi\left[\frac{\dfrac{\mu_0-a}{\sigma} + \dfrac{a-\mu_1}{\sigma}}{2}\right].$$

That is,

$$R(\mathbf{d},a^*(\mathbf{d})) = \Phi\left[\frac{\Phi^{-1}(p_0) + \Phi^{-1}(p_1)}{2}\right]. \tag{16}$$

The function $P(\mathbf{d}) = \min_a R(\mathbf{d},a) = R(\mathbf{d},a^*(\mathbf{d}))$ is a PerMIA, because it can be used to determine the value of the nonadjustment design parameters $\mathbf{d}$, at which channel performance is optimal while ignoring the value of the adjustment parameter a.

It follows that the engineer can follow the two-step procedure given below to find the optimal value $(\mathbf{d}^*, a^*)$ of $(\mathbf{d}, a)$:

Procedure 6.
Step 1. Find d^* that minimizes

$$\Phi \left[\frac{\Phi^{-1}(p_o(\mathbf{d}, a)) + \Phi^{-1}(p_1(\mathbf{d}, a))}{2} \right],$$

where a is in the arbitrary present position.

Step 2. Find a^* for which

$$p_0(\mathbf{d}^*, a^*) = p_1(\mathbf{d}^*, a^*).$$

Step 2 follows because the misclassification probabilities are equal if and only if a minimizes $R(\mathbf{d}^*, a)$. Step 2 is carried out by empirically adjusting a back and forth until misclassification probabilities are equal.

We note that without Procedure 6, finding the optimal value $\mathbf{d}^*$ would involve doing the above time-consuming adjustment of a for each candidate value of $\mathbf{d}$.

12.5.2 EXPLANATION OF TAGUCHI'S SN RATIO

Taguchi and Phadke[3] (following Taguchi[2]) suggested a formula other than (16) for their two step procedure. We have found that their formula follows from the preceding argument if the error ε caused by noise has a standard logistic distribution rather than a standard normal distribution. Taguchi and Phadke did not use that value as the performance measure, however. Instead, they use a decreasing function of it that they call the SN ratio. In Appendix B we show that this SN ratio would be a meaningful measure if the binary channel problem were modified so that the received voltages were observable. The reader interested in understanding the relevance of the SN ratio concept of communication engineering to parameters design may want to read Appendix B.

12.6 ADJUSTMENT PARAMETERS IN PRACTICE

12.6.1 HOW ADJUSTMENT PARAMETERS ARISE IN PRACTICE

The problem of identifying adjustment design parameters may seem difficult, but in practice these parameters are often suggested by engineering knowledge and product or process design conventions. In particular, adjustment parameters are frequently those design parameters used to fine tune performance after the other, more difficult to change, design parameters have been set.

As illustrated previously, examples of adjustment parameters include the size of the mold used in making tiles, scale markings on a measuring instrument, and the gear ratio in a steering mechanism (Appendix A). Other examples of adjustment parameters are the length of the pendulum in a clock and the treble, base, volume and balance knobs on a stereo receiver.

12.6.2 THE VALUE OF ADJUSTMENT PARAMETERS FOR PRODUCT AND PROCESS DESIGNERS

Adjustment parameters are often specifically designed into a product or process to make the design more flexible. For example, certain changes in design specifications may be accommodated by changing the setting of an adjustment parameter. In the tile example, a change in the desired tile size is easily accomplished by changing the mold size. The clay formulation need not be changed.

In other examples, adjustment parameters may make a product adaptable to a variety of customer use conditions. The carpet height adjustment of a vacuum cleaner, for example, allows good cleaning on a range of carpet pile depths.

12.6.3 ADVANTAGES OF MEASURING PERFORMANCE INDEPENDENT OF ADJUSTMENT PARAMETERS

When adjustment parameters are included in the design of a product or process, measuring that product or process's performance independent of adjustment parameter settings can be very advantageous. For example, when an adjustment parameter is meant to accommodate future changes in design specifications (such as the desired tile size), use of a PerMIA as performance measure ensures that nonadjustment design parameter values that are optimal for one specification are also optimal for other specifications.

If the adjustment is going to be made by the customer, a PerMIA can be particularly useful to a product designer. It allows the designer to gauge the performance of the product independently of adjustments the customer might make to suit his particular situation. For the product designer, the PerMIA measures performance *as it would be* after proper adjustment. There is no need to test each alternative product design under every customer-use condition.

Use of a PerMIA also reduces the dimensionality of the parameter design optimization problem. This may be advantageous since lower dimensional optimization problems are generally easier to solve.

A PerMIA can sometimes be used to turn a constrained parameter design optimization problem into an unconstrained one, as is the case in the static parameter design problem (see the discussion following Procedure 3).

PerMIAs can be used to simplify parameter design experiments. PerMIAs can be derived *analytically* using minimal engineering knowledge of product or process behavior (as captured by the model). The empirical model identification and fitting is limited to $P(\mathbf{d})$ and $R(\mathbf{d}^*, a)$ rather than the more complex $R(\mathbf{d}, a)$ which could require cross-terms between the parameters $\mathbf{d}$ and $\mathbf{a}$.

12.6.4 EMPIRICAL METHODS FOR IDENTIFYING A PerMIA

Choice of a performance measure independent of adjustment depends upon some knowledge of the form of the transfer-function model. In a particular application, engineering knowledge might indicate a multiplicative model like (2) rather than an additive one like (7). Empirical studies can check or supplement engineering knowledge, however. For the static case, Box[5] has described how the form of the transfer function model can be investigated by estimating the transformation of the output Y required to induce additivity in the transfer-function model.

12.7 SUMMARY

Parameter design is a process for finding product designs that are insensitive to noise such as manufacturing and environmental variability. Operationally, the objective of parameter design is to find the setting of the product's design parameters that minimizes expected loss due to noise. G. Taguchi pioneered the use of statistically planned experiments for parameter design. His work has led to wide application of this method in Japan and its increasing use by U.S. industry.

In parameter design, Taguchi (see Taguchi;[2] Taguchi and Phadke[3]) used performance measures that he called SN ratios. In general, he gave no connection between these performance measures and his stated objective of minimizing loss caused by noise.

In this article we show that *if certain models for the product or process response are assumed*, then maximization of the SN ratio leads to minimization of average squared-error loss. The SN ratios take advantage of the existence of special design parameters, called adjustment parameters. When these parameters exist, use of the SN ratio allows the parameter design optimization procedure to be conveniently decomposed into two smaller optimization steps.

In these situations the desire to find a two step optimization procedure seems to be a major motivation behind Taguchi's use of SN ratios. When different models hold, however, a two-step procedure is possible only if a performance measure different from the SN ratio is used. Hence there are many real problems in which the SN ratio is not independent of adjustment parameters, and its use could lead to far from optimal design parameter settings.

Because adjustment parameters bring definite advantages to product and process designers, we propose a type of performance measure that takes advantage of the existence of adjustment parameters and is more general than Taguchi's SN ratios. We call these measures performance measures independent of adjustment, or PerMIAs.

When an adjustment parameter exists, a PerMIA can be conveniently derived directly from knowledge of the loss function and the general form of the transfer-function model. We have given a procedure for doing this and illustrated its use in several generic parameter design problems. In some of these problems we saw that following the procedure leads directly to the SN ratio Taguchi recommended.

ACKNOWLEDGMENTS

The authors gratefully acknowledge the helpful comments of R. Berk, K. Dehnad, R. Easterling, C. Fung, A. B. Godfrey, J. H. Hooper, W. Q. Meeker, V. N. Nair, W. A. Nazaret, P. G. Sherry, C. S. Sherrerd, C.F. Wu and the referees.

[Discussions of this article and the author's response appear in *Technometrics* vol. 29, no. 3 (August 1987):266-285.]

APPENDIX A

DESIGNING A TRUCK STEERING MECHANISM, A DETAILED EXAMPLE

In this section we show how a PerMIA is derived in a realistic control parameter design problem, the design of a truck steering mechanism. This example, a modification of one given by Taguchi and Wu,[1] also illustrates how adjustment parameters arise in practice. The truck driver chooses the steering angle s to make a turn of radius t. (Turning angle may seem a more natural measure of the truck's response to a certain steering angle, but turning radius is much easier to measure.) The chosen steering angle is determined by the driver's control strategy. In designing the truck-steering mechanism, the engineer's objective is to minimize the expected loss caused by deviation of the truck's actual turning radius from the driver's intended turning radius. This deviation is caused by noise such as road condition, load position, and tire pressure. (Driving speed is another noise which has a profound effect on how the truck responds to a given steering angle. For simplicity, we assume that speed is constant. See Taguchi and Wu[1] for a solution that includes speed as a noise factor.)

As mentioned previously, one way to make the steering mechanism insensitive to noise is to weld the steering mechanism so that the truck always goes straight. Of course, this would not do, because, in addition to minimizing sensitivity to noise, the design must allow the driver to make any needed turn using a comfortable steering angle. In particular, since the size of the steering angle is inversely related to the force required to turn the steering wheel, comfortable steering angles are neither too large nor too small—small angles require too much exertion; large angles require too many turns of the steering wheel. Suppose the engineer expects that the driver will need to make turns between t_L meters and t_H meters and comfortable steering angles for making these turns are between s_L and s_H degrees.

As before, the block diagram in Figure 12.3 summarizes the problem from the point of view of parameter design. The truck driver who intends to make a turn of radius t chooses a steering angle s using his control strategy $s(t)$. The truck responds with a turn of radius Y, which depends on the steering angle s, the noise conditions N, and the design of the steering mechanism.

In this example the design of the steering mechanism is determined by three design parameters: hardness of front springs (A), type of steering geometry (B), and gear ratio (G). From the geometry of steering, it can be shown that it is reasonable to assume that the relationship between Y, s, N, A, B, and G is given by the transfer function:

$$Y = (Gs)^{-\delta(N,A,B)} \tag{17}$$

where δ is a positive function.

Taking logs and doing some obvious renaming, the transfer function can be rewritten as:

$$\log Y = \left[\alpha(A,B,G) - \beta(A,B)\log s \right] \varepsilon(N,A,B) \tag{18}$$

where we force the condition

$$E_N(\varepsilon(N,A,B)) = 1 .$$

Model (18) is similar to model (14) of Section 12.4.2, except that the error term is now multiplicative rather than additive. Both models, however, lead to the same PerMIA, as shown in the following.

Recall that the designer's objective is to find the setting of the design parameters that minimizes the expected loss caused by deviation of Y from t. If this loss is $L(Y,t) = (Y-t)^2$, then no PerMIA seems to be available. A PerMIA can be obtained, however, if the loss could be well-approximated by

$$L(Y;t) = (\log Y - \log t)^2 .$$

Then expected loss is

$$R(A,B,G) = E_s E_N \left[(\log Y - \log t)^2 \mid A,B,G,s(t) \right]$$

$$= E_s \left\{ Var_N(\log Y \mid A,B,G,t) + \left[E_N(\log Y \mid A,B,G,t) - \log t \right]^2 \right\} .$$

Hence, assuming the control strategy

$$s(t) = \left\{ s\colon \; E_N \left[\log Y \mid A,B,G,t \right] = \log t \right\},$$

under Model (18), we get

$$R(A,B,G) = E_s \left[(\log t)^2 \right] \sigma^2(A,B) ,$$

where

$$Var_N(\varepsilon(N,A,B)) = \sigma^2(A,B).$$

Therefore, the ability of a driver to make a desired turn is measured by

$$\sigma^2(A,B) . \tag{19}$$

It is not enough to find the values of A and B to minimize $\sigma^2(A,B)$, however. Recall that the steering mechanism must allow the driver to make any turn of radius between t_L meters and t_H meters using steering angles between s_L and s_H degrees. From Model (18), this will be possible on the average if the slope coefficient satisfies the constraint

$$\beta^2(A,B) \geq \left[\frac{\log t_L - \log t_H}{\log s_H - \log s_L} \right]^2 . \tag{20}$$

If a design satisfies this constraint, the gear ratio, G, can be adjusted so that

$$E(\log Y \mid A,B,G,s_H) \leq \log t_L < \log t_H \leq E(\log Y \mid A,B,G,s_L) .$$

Notice that neither the performance measure $\sigma^2(A,B)$ nor the constraint (14) depends on the gear ratio G. That is, $\sigma^2(A,B)$ is a PerMIA, and its constrained optimization can be done with no dependence on the value of the adjustment parameter G.

Notice that with model (18) it is not even clear how to substitute into Taguchi's SN ratio formula (12) for the continuous-continuous dynamic parameter design problem.

APPENDIX B

TAGUCHI'S SN FOR THE BINARY CHANNEL EXAMPLE

Communication engineers often use SN ratio as a measure of a transmission channel's performance. This is done because in many types of channels the probability of transmission error is a decreasing function of the SN ratio, or the channel capacity is an increasing function of the SN ratio. The channel capacity is the theoretical maximum bit rate of information that can be transmitted over the channel. (See Chapters 8 and 9 of Pierce.[6])

Assume that in Section 12.5 the received voltages $\mu_0 + \sigma\varepsilon$ and $\mu_1 + \sigma\varepsilon$ are observable. Let $\mu = \dfrac{\mu_0 + \mu_1}{2}$, $\alpha_0 = \mu_0 - \mu$, $\alpha_1 = \mu_1 - \mu$, and $\varepsilon\prime = \sigma\varepsilon$. Then the received voltages $X_i (i = 1, 2, ...)$ have the form

$$X_{i,k} = \mu + \alpha_k + \varepsilon\prime_{i,k}, \quad k = 0, 1$$

$$\alpha_0 + \alpha_1 = 0$$

$$\varepsilon_i^\prime \ wig \ N(0, \sigma^2) . \tag{21}$$

The variance σ^2 is the average power of the noise and

$$\sigma_\alpha^2 = \frac{\alpha_0^2 + \alpha_1^2}{2} = \alpha_0^2 = \alpha_1^2$$

is the average power of the signal. (Since power is voltage times current and by Ohm's law voltage is proportional to current, it follows that power is proportional to voltage squared.) Hence σ_α^2/σ^2 is the ratio of average signal power to average noise power (SN ratio), and the quantity

$$10 \ log_{10}(\sigma_\alpha^2/\sigma^2)$$

is how strong the signal power is relative to the noise power in decibel units.

Now as shown by Chapter 8 of Pierce and Posner[7] the transmission error probability after leveling is a decreasing function of the SN ratio. Hence an engineer trying to find the settings of design variables that minimize this probability could pick the settings that maximize the SN ratio, or equivalently, its decibel value. In other words, the SN ratio is an appropriate performance measure.

The problem is how to estimate the SN ratio on the basis of sending a known sequence of n zeros and n ones through the channel. One can think of model (21) as a one-way layout analysis of variance problem. Then we see that the expected mean squared for treatment (MST) is given by

$$E(MST) = \sigma^2 + n(2\sigma_\alpha^2),$$

and the expected mean squared for error (MSE) is given by

$$E(MSE) = \sigma^2.$$

Hence a simple estimate of the SN ratio σ_α^2/σ^2 is the method of moments estimate, given by

$$\frac{(MST-MSE)}{2nMSE}.$$

So the engineer conducting the experiment would try to find the setting of design variables that maximizes

$$10 \ log_{10}\{(MST-MSE)/(nMSE)\} . \tag{22}$$

In the original Binary Channel problem, however, voltage is not observable. In solving this problem, Taguchi and Phadke[3] suggested that the estimate (22) still be used on the binary output. The binary output is used to calculate MST and MSE, which are then substituted into formula (22). After some algebra, this process yields

$$\eta = 10 \ log_{10}\left\{\frac{(1-p_0-p_1)^2}{p_0(1-p_0)+p_1(1-p_1)}\right\} . \tag{23}$$

Then they used a formula which is equivalent to formula (16) if ϕ is the standard logistic density instead of the standard normal density. Upon substitution of this formula in formula (23), they get what they called the SN ratio after leveling:

$$\eta = 10 \ log_{10} \frac{(1-2q)^2}{2q(1-q)} . \tag{24}$$

where q is the counterpart of $R(\mathbf{d}, a^*(\mathbf{d}))$ in formula (16) for the logistic distribution.

Unfortunately, Model (21) cannot hold when X_i, k ($i = 1, 2, \cdots, n$; $k = 1, 2$) are binary variables. Moreover, since these binary variables are not voltages (or currents), it is not clear in what sense η is a decreasing function of q on $[0, \frac{1}{2}]$, since after some algebra it can be shown that

$$\eta = 10 \, log_{10} \left[\frac{8}{\frac{1}{(q-.5)^2} - 4} \right].$$

Thus, the setting of the design parameters that maximizes η also minimizes q. It follows that the approach of Taguchi and Phadke[3] is equivalent to that given in Section 12.5.1.

REFERENCES

1. Taguchi, G., and Y. Wu. 1980. *Introduction to Off-line Quality Control.* Nagaya, Japan: Central Japan Quality Control Association. (Available from the authors through American Supplier Institute, 32100 Detroit Industrial Expressway, Romulus, MI, 48174.) 8.

2. Taguchi, G. 1976, 1977. *Experimental Design.* Vols. 1 and 2. Tokyo: Maruzen Publishing Company. (Japanese)

3. Taguchi, G., and M. S. Phadke. 1984. "Quality Engineering Through Design Optimization." *Proceedings of GLOBECOM 84 Meeting,* 1106-1113. IEEE Communication Society.

   (Quality Control, Robust Design, and the Taguchi Method; Article 5)

4. Phadke, M. S. 1982. "Quality Engineering Using Design of Experiments." *Proceedings of the American Statistical Association, Section on Statistical Education,* 11-20. Cincinnati. (Aug.)

   (Quality Control, Robust Design, and the Taguchi Method; Article 3)

5. Box, G. E. P. 1986. "Studies in Quality Improvement: Signal-to-Noise Ratios, Performance Criteria and Statistical Analysis: Part I." *Technical Report No. 11.* Madison, Wis.: Center for Quality and Productivity Improvement, University of Wisconsin.

6. Pierce, J. R. 1980. *An Introduction to Information Theory: Symbols, Signals, and Noise.* 2d ed., rev. New York: Dover Publications, Inc.

7. Pierce, J. R., and E. C. Posner. 1980. *Introduction to Communication Science and Systems.* New York and London: Plenum Press.

ADDITIONAL REFERENCES

Kackar, R. N. 1985. Off-line Quality Control, Parameter design and the Taguchi Method. *Journal of Quality Technology* **17** (Oct. no. 4):176-209.

   (Quality Control, Robust Design, and the Taguchi Method; Article 4)

McGraw-Hill. 1977. *McGraw-Hill Encyclopedia of Science and Technology.* Vol. 3. New York: McGraw-Hill, Inc.

13

A GEOMETRIC INTERPRETATION OF
TAGUCHI'S SIGNAL TO NOISE RATIO

Khosrow Dehnad

13.1 INTRODUCTION

The Taguchi Method is an empirical way of improving the quality of a product or process (a system). It consists of experimentally determining the levels of the system's parameters so that the system is insensitive (robust) to factors that make it's performance degrade, Phadke & Dehnad.[1] The successful application of this method to various engineering problems has generated an interest among statisticians in studying the methodology and casting it in a traditional statistical framework. A concept of particular interest has been the notion of "Signal to Noise Ratio" (S/N ratio) that Professor Taguchi uses to measure the contribution of a system parameter[1] to variability of the functional characteristics of that system. Moreover, he uses the S/N ratio in a two step optimization procedure that aims at reducing the variability of these functional characteristics by the proper setting of these parameters. The first step consists of conducting highly fractional factorial experiments (Orthogonal Array Designs) at various levels of the system's parameters and using the data to select those levels that maximize the S/N ratio. This reduces the variability of the functional characteristic of the system. However, this characteristic might have some deviation from its desired value and the next step is to eliminate such deviations through proper adjustment of a system's parameter called the "adjustment factor."[2]

---

1. Glossary equivalent is *control parameter*.
2. Glossary equivalent is *adjustment parameter*.

Taguchi does not give a formal definition of the S/N ratio and a rule for its selection. There have been attempts by Phadke & Dehnad,[1] Kackar,[2] and Leon, Shoemaker & Kackar[3] to formally define and interpret this notion; also, Taguchi & Phadke[4] divide the engineering problems into general categories and suggest a S/N ratio for each category.

In this paper we first use the notion of quadratic loss function to arrive at a formal definition of S/N ratio as a measure of the angular deviation of the functional characteristic of a system from its target. In Section 13.3, we consider the two step optimization procedure based on the notions of S/N ratio and an "adjustment factor" and give a geometric interpretation of it. We also show that this procedure actually minimizes the expected loss. In Section 13.4, we use a case study to illustrate the results of the previous sections. Finally, in Section 13.5 by considering the example of a communication channel we argue that blind application of the methodology could be misleading and in certain cases quite inappropriate. Details of some mathematical results used in the paper are presented in the appendix.

13.2 SIGNAL TO NOISE RATIO

In this section we formally define the notion of signal to noise ratio as a measure of angular deviation of the functional characteristic of a system from its target. First we introduce some notation and terminology that will be used in the text.

Following Taguchi & Phadke,[4] we represent a system (i.e., a product or a process) by the P-Diagram in Figure 13.1.

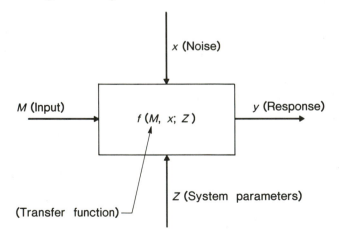

Figure 13.1. P-Diagram of a Product/Process

In this diagram, M is the input,[3] y is the output, and z is the vector of control factors[4] whose levels are set by the designer; f represents the transfer function and x is the noise, i.e., the causes of possible deviation of y from its target y_0. For further discussion of these concepts see Taguchi & Phadke[4] and Leon and others.[3] Any deviation of y from its desired value y_0 results in a loss of quality and in many cases this loss can be reasonably approximated by a quadratic function

$$l(y) = k \ (y - y_0)^2.$$

For a justification of this and a way of determining k based on economic considerations see Taguchi[5] and Phadke & Dehnad.[1] It frequently happens that k is a function of M the input; for example, in the case of a thermostat of a refrigerator the loss due to deviation of temperature from the setting of the thermostat could depend on the selected temperature as well as the magnitude of deviation. Because relatively small deviation from certain temperatures can translate into spoiling of food and possibly considerable loss.

In cases where k depends on the input M the expected loss for a given configuration "z" of the system and an input M is

$$Q(M) = \int k(M) \ \Big[f(M,x \ ; z) - g_0(M) \Big]^2 dP_1(x) \tag{1}$$

where $P_1(.)$ is the distribution of the noise and $k(M) \geq 0$ is the coefficient of the quadratic loss function. The total expected loss is obtained by summing (1) with respect to the frequency of various input values. If $P_2(.)$ is the distribution of M, the input values, then the expected loss is

$$\underset{M,x}{\overline{Q}} = \int Q(M) \ dP_2(M) = || \ f(z) - g_0 \ ||^2.$$

This quantity has the property of squared distance between the functions f and g_0. Further, we can formally define the angle θ between f and g_0 by the following relation (Figure 13.2 and appendix)

$$Cos(\theta) = \frac{\int k(M) \ (f \cdot g) \ dP_1(x) \ dP_2(M)}{|| f || \ \ || g_0 ||}.$$

---

3. Glossary equivalent is *signal parameter.*
4. Glossary equivalent is *control parameter.*

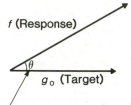

f (Response)

θ

*g*₀ (Target)

Angular deviation of
response from target

Figure 13.2. Angular Deviation of Response

Note the distribution of noise $P_1(.)$ is assumed to be independent of the input signal M. This assumption in the case of a thermostat translates into the fact that noise factors such as the level of humidity do not depend on the setting of the thermostat.

In this geometry since

$$g_0 - \frac{||g_0||}{||f||} Cos(\theta) f$$

is perpendicular to f, it follows from the Pythagoras theorem that expected loss can be written as the sum of two components (Figure 13.3)

$$||f - g_0||^2 = ||f - \frac{||g_0||}{||f||} Cos(\theta) f||^2 + ||g_0 - \frac{||g_0||}{||f||} Cos(\theta) f||^2.$$

Of these two components,

$$||f - \frac{||g_0||}{||f||} Cos(\theta) f||^2,$$

which will be referred to as the adjustable component of loss, can usually be corrected easily. For example, in the case of the thermostat if the actual temperature is always twice its selected value, a calibration easily corrects this problem. On the other hand,

$$||g_0 - \frac{||g_0||}{||f||} Cos(\theta) f||^2$$

is generally hard to adjust and is referred to as the non-adjustable component of the loss. Note the loss after elimination of its systematic part becomes

$$|\,|\,g_0\,|\,|^2\,Sin^2(\theta)$$

and in this case minimizing the loss is equivalent to minimizing the angle θ.

This angle can be viewed as the angular deviation of response f from its target g_0, and a decreasing function of θ is called a signal to noise ratio.

In practice the choice of a particular S/N ratio depends on such considerations as the desired range of S/N ratio or its additivity with respect to certain system parameters. One commonly used measure is $Log\ Cot^2(\theta)$ which has the real line as its range.

Further, $Cot\ (\theta)$ is a measure of relative magnitude of the desirable part of the output "signal" to the undesirable part "noise" (Figure 13.4). In particular, when the target has a fixed value i.e., $g_0 = \mu_0$ then

$$Cos^2(\theta) = \frac{\left[E[\mu_0\ f\,]\right]^2}{E[\mu_0^2]\ E[\ f^2\ ]}$$

$$= \frac{\mu_0^2\ E[f\,]^2}{\mu_0^2\ E[\ f^2\ ]} = \frac{E[f\,]^2}{Var[f\,] + E[f\,]^2}$$

$$= \frac{1}{1 + Var[f\,]\,/\,E[f\,]^2}\ .$$

We also have

$$Cos^2(\theta) = \frac{Cot^2(\theta)}{Cot^2(\theta) + 1}\ ,$$

therefore,

$$Cot^2(\theta) = \frac{E[f\,]^2}{Var[f\,]}\ .$$

So if the S/N ratio is an increasing function of $Cot\theta$, then maximizing it is equivalent to minimizing the coefficient of variation.

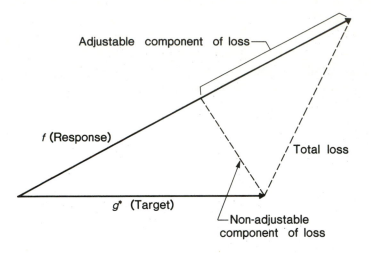

Adjustable component of loss

f (Response)

Total loss

g* (Target)

Non-adjustable
component of loss

Figure 13.3. Decomposition of Loss

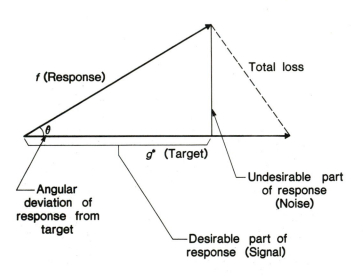

f (Response)

Total loss

θ

g* (Target)

Angular
deviation of
response from
target

Undesirable part
of response
(Noise)

Desirable part of
response (Signal)

Figure 13.4. Decomposition of Response

13.3 THE SCALING FACTOR AND A TWO STEP OPTIMIZATION PROCEDURE

One way of improving the quality of a system, i.e., reducing $||f - g_0||^2$ the expected loss, is through proper settings of the system's parameters. To make the system insensitive to external factors (noise) is referred to as "Robust System Design" and can be accomplished analytically if f, the transfer function, is known. In the absence of this knowledge, experimentation can be used to empirically arrive at better settings of the process parameters.

When choosing the value of a parameter with the goal of reducing the loss, we should consider how this choice affects both the non-adjustable and the adjustable components of the loss, and this could be a formidable task. However, if there is a control parameter z_1, called the adjustment factor, that satisfies

$$f(x,M; z_1,z) = f_1(z_1) f_2(x,M; z)$$

then we can eliminate the above need of having to simultaneously consider the effect of our action on the two components of the loss. Because in this case we can first select the parameter levels that maximize the S/N ratio. This minimizes the angular deviation of the system from its target and consequently the non-adjustable component of the loss (Figure 13.2). Moreover, while doing this we need not be concerned about the effect of our action on the adjustable component of the loss since we can always eliminate this component by the proper setting of the adjustment factor. Further, such adjustments will not affect the S/N ratio hence the non-adjustable component of the loss. To see this note that

$$Cos^2(\theta) = \frac{\left[\underset{\scriptscriptstyle M}{E} \underset{\scriptscriptstyle x,M}{g_0 f} \right]^2}{\underset{M}{E} g_0^2 \left[\underset{M,x}{E} f^2 \right]}$$

$$= \frac{f_1^2 \left[\underset{M,x}{E} g_0 f_2 \right]^2}{f_1^2 \underset{M}{E} \left[g_0 \right]^2 \left[\underset{M,x}{E} f_2^2 \right]} = \frac{\left[\underset{M,x}{E} g_0 f_2 \right]^2}{\underset{M}{E} \left[g_0 \right]^2 \left[\underset{M,x}{E} f_2^2 \right]} .$$

This implies that $Cos(\theta)$ hence θ do not depend on z_1, so varying this factor does not affect the angular deviation of the response from its target. Therefore, we can use this factor to eliminate any adjustable deviation of the response from the target without affecting the non-adjustable component of the loss which, by the way, has already been minimized.

In short, the above two step procedure consists of

a) Determining the levels of control factors that maximize the S/N ratio. This minimizes the angular deviation of the response from its target hence the non-adjustable component of the loss.

b) Using the adjustment factor(s) to eliminate any adjustable deviation of the response from the target.

The above procedure greatly simplifies the task of arriving at improved settings of the process parameters by eliminating the need for simultaneously considering the two components of the loss when choosing the new settings of the system parameters. Also, if there is a change in the scale of the target, there is no need to reconfigure all the system's parameters because such changes do not affect the angular deviation of the response, hence the non-adjustable component of the loss, and there is no need to readjust the parameters that affect this component of the loss (Figure 13.3). The above change only affects the adjustable component of the loss, and this loss can be easily be taken care of through the use of adjustment factor.

In practice if adjustment factors are not known beforehand, data analysis, graphical and statistical techniques can be used to identify them. Further, if there is more than one adjustment factor the choice of one is usually based on practical considerations such as ease of change. For example, in the case of window photolithography (Phadke and others,[6]) which is considered in the next section, the viscosity was discovered to be an adjustment factor; however, this factor was not easy to adjust so it was passed over in favor of exposure which was easier to control.

Clearly, the above discussions are based on the assumption of a quadratic loss function. A generalization of the concept of signal to noise ratio and the resulting two step optimization is given in Leon, Shoemaker, and Kackar.[3]

13.4 A CASE STUDY

To illustrate the discussions of the previous sections, let us consider the following study by Phadke, et al.[6] The goal of the study was to improve the yield of an integrated circuit fabrication process. Here, we only consider the part of the study that deals with improving the quality of pre-etch line width, i.e., reducing its variability while keeping the average width on target. The reader is referred to the original paper for more details.

The factors identified as system parameters and their levels are given in Table 13.1.

TABLE 13.1. TEST LEVELS

Labels	Factors Name	Levels		
			Standard Levels	
A	Mask Dimension (μm)		2	2.5
B	Viscosity		204	206
C	Spin Speed (rpm)	Low	Normal	High
D	Bake Temperature (°C)	90	105	
E	Bake Time (min)	20	30	40
F	Aperture	1	2	3
G	Exposure Time	20% Over Normal	20% Under	
H	Developing Time (s)	30	45	60
I	Plasma Etch Time (min)	14.5	13.2	15.8

Dependence of spin speed on viscosity

		Spin Speed (rpm)		
		Low	Normal	High
Viscosity	204	2000	3000	4000
	206	3000	4000	5000

Dependence of exposure on aperture

		Exposure (PEP-Setting)		
		20% Over	Normal	20% Under
Aperture	1	96	120	144
	2	72	90	108
	3	40	50	60

Eighteen experiments were conducted according to Table 13.2.

Each experiment involved two wafers, and five measurements of the pre-etch line width were made on each wafer, Table 13.3.

TABLE 13.2. THE L_{18} ORTHOGONAL ARRAY

Experiment Number	Column Number and Factor							
	1 A	2 BD	3 C	4 E	5 F	6 G	7 H	8 I
1	1	1	1	1	1	1	1	1
2	1	1	2	2	2	2	2	2
3	1	1	3	3	3	3	3	3
4	1	2	1	1	2	2	3	3
5	1	2	2	2	3	3	1	1
6	1	2	3	3	1	1	2	2
7	1	3	1	2	1	3	2	3
8	1	3	2	3	2	1	3	1
9	1	3	3	1	3	2	1	2
10	2	1	1	3	3	2	2	1
11	2	1	2	1	1	3	3	2
12	2	1	3	2	2	1	1	3
13	2	2	1	2	3	1	3	2
14	2	2	2	3	1	2	1	3
15	2	2	3	1	2	3	2	1
16	2	3	1	3	2	3	1	2
17	2	3	2	1	3	1	2	3
18	2	3	3	2	1	2	3	1

TABLE 13.3. EXPERIMENTAL DATA

Experiment Number	Line-Width Control Feature Photoresis-Nanoline Tool (Micrometers)					
	Top	Center	Bottom	Left	Right	Comments
1	2.43	2.52	2.63	2.52	2.5	
1	2.36	2.5	2.62	2.43	2.49	
2	2.76	2.66	2.74	2.6	2.53	
2	2.66	2.73	2.95	2.57	2.64	
3	2.82	2.71	2.78	2.55	2.36	
3	2.76	2.67	2.9	2.62	2.43	
4	2.02	2.06	2.21	1.98	2.13	
4	1.85	1.66	2.07	1.81	1.83	
5	-	-	-	-	-	Wafer Broke
5	1.87	1.78	2.07	1.8	1.83	

TABLE 13.3.   EXPERIMENTAL DATA
(Continued)

Experiment Number	Line-Width Control Feature Photoresis-Nanoline Tool (Micrometers)					
	Top	Center	Bottom	Left	Right	Comments
6	2.51	2.56	2.55	2.45	2.53	
6	2.68	2.6	2.85	2.55	2.56	
7	1.99	1.99	2.11	1.99	2.0	
7	1.96	2.2	2.04	2.01	2.03	
8	3.15	3.44	3.67	3.09	3.06	
8	3.27	3.29	3.49	3.02	3.19	
9	3.0	2.91	3.07	2.66	2.74	
9	2.73	2.79	3.0	2.69	2.7	
10	2.69	2.5	2.51	2.46	2.4	
10	2.75	2.73	2.75	2.78	3.03	
11	3.2	3.19	3.32	3.2	3.15	
11	3.07	3.14	3.14	3.13	3.12	
12	3.21	3.32	3.33	3.23	3.10	
12	3.48	3.44	3.49	3.25	3.38	
13	2.6	2.56	2.62	2.55	2.56	
13	2.53	2.49	2.79	2.5	2.56	
14	2.18	2.2	2.45	2.22	2.32	
14	2.33	2.2	2.41	2.37	2.38	
15	2.45	2.50	2.51	2.43	2.43	
15	-	-	-	-	-	No wafer
16	2.67	2.53	2.72	2.7	2.6	
16	2.76	2.67	2.73	2.69	2.6	
17	3.31	3.3	3.44	3.12	3.14	
17	3.12	2.97	3.18	3.03	2.95	
18	3.46	3.49	3.5	3.45	3.57	
18	-	-	-	-	-	No wafer

The Signal to Noise Ratio was

$$S/N = \text{Log} \frac{\text{mean line width}}{\text{Standard Deviation of the line width}}.$$

The mean line width and S/N ratio corresponding to each run are given in Table 13.4.

Under the assumption of linearity of S/N ratio with respect to design parameters, the ratio for different levels of the design parameters are given in Table 13.5.

TABLE 13.4. PRE-ETCH LINE-WIDTH DATA

Experiment Number	Mean Line Width, $\bar{\chi}$ (μm)	Standard Deviation of Line Width, s (μm)	s/n $\eta = log(\bar{\chi}/s)$
1	2.500	0.0827	1.4803
2	2.684	0.1196	1.3512
3	2.660	0.1722	1.1889
4	1.962	0.1696	1.0632
5	1.870	0.1168	1.2043
6	2.584	0.1106	1.3686
7	2.032	0.0718	1.4520
8	3.267	0.2101	1.1917
9	2.829	0.1516	1.2709
10	2.660	0.1912	1.1434
·11	3.166	0.0674	1.6721
12	3.323	0.1274	1.4165
13	2.576	0.0850	1.4815
14	2.308	0.0964	1.3788
15	2.464	0.0385	1.8065
16	2.667	0.0706	1.5775
17	3.156	0.1569	1.3036
18	3.494	0.0473	1.8692

TABLE 13.5. PRE-ETCH LINE WIDTH FOR AVERAGE S/N

	Factor	Average s/n		
		Level 1	Level 2	Level 3
A	Mask Dimension	1.2857	1.5166	
BD	Viscosity Bake Temperature	(B_1D_1) 1.3754	(B_2D_1)11.3838	(B_1D_2) 1.4442
B	Viscosity	1.4098	1.3838	
D	Bake Temperature	1.3796	1.4442	
C	Spin Speed	1.3663	1.3503	1.4868
E	Bake Time	1.4328	1.4625	1.3082
F	Aperture	1.5368	1.4011	1.2654
G	Exposure Time	1.3737	1.3461	1.4836
H	Developing Time	1.3881	1.4042	1.4111

Overall average s/n = 1.4011.

The average line width for different levels of design parameters are given in Table 13.6.

TABLE 13.6. PRE-ETCH LINE WIDTH FOR THE MEAN LINE WIDTH

Factor	Mean Line Width (μm)		
	Level 1	Level 2	Level 3
A Mask Dimension	2.39	2.87	
BD Viscosity Bake Temperature	(B_1D_1) 2.83	(B_2D_1)2.31	(B_1D_2) 2.74
B Viscosity	2.79	2.31	
D Bake Temperature	2.57	2.74	
C Spin Speed	2.40	2.59	2.89
E Bake Time	2.68	2.68	2.53
F Aperture	2.68	2.56	2.64
G Exposure Time	2.74	2.66	2.49
H Developing Time	2.60	2.60	2.69

Overall mean line width = 2.63 μm.

Expressing the previous table in terms of the angular deviation of the response from its target we get the values shown in Table 13.7.

TABLE 13.7. ANGULAR DEVIATION OF RESPONSE FROM TARGET

Factor	Average 20θ		
	Level 1	Level 2	Level 3
A Mask Dimension	60	34	
BD Viscosity Bake Temperature	48	48	42
B Viscosity	44	48	
D Bake Temperature	48	42	
C Spin Speed	50	52	38
E Bake Time	42	40	56
F Aperture	34	46	62
G Exposure Time	48	52	38
H Developing Time	46	46	44

To determine possible scaling factors, let us graphically represent Tables 13.6 and 13.7 by Figure 13.5. In Figure 13.5, a dotted line stands for the first level, the broken line for the second level, and the solid line for the third level (when applicable) of a design parameter. The length of the line corresponds to the mean effect and the angle corresponds to the angular deviation (20θ) of the response from the target.

It can be seen from the graphs that factors A and F have pronounced effect on the angular deviation of response (hence the non-adjustable component of loss) and factor B seems to be a good candidate for the scaling factor. However, in practice factor B was passed over in favor of factor C because viscosity was hard to control and adjust.

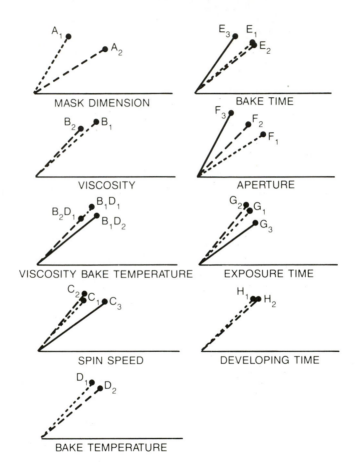

Figure 13.5. Mean and Angular Deviation of Response from Target

13.5 COMMUNICATION CHANNEL

A major step in applying the optimization procedure of the previous section is to formulate our problem in such a way that the method can be applied to it. To do this, we have to use engineering knowledge and data analysis; because we can not simply view an engineering problem as a black-box and hope to solve it by applying a method that does not consider the inner working of the system and is solely based on the characteristics of the elements of its P-Diagram, e.g., whether the input is discrete or the output is continuous with a fixed target and so on. For example in the case of optimizing the design of a communication channel we can-not assign numerical values to input and output signals and use the above pro-cedure to optimize the design; because, in this case the loss can not be approxi-mated by a quadratic function of the output. To be more precise a communication channel can be represented by the sets X and Y of transmitted and received signals

$$X = \left\{ x_i \mid i=1,2..I \right\}$$

$$Y = \left\{ y_j \mid j=1,2......J \right\}$$

and the $I \times J$ matrix (p_{ij}) with p_{ij} the probability of transmitting x_i and receiving y_j. For this channel the capacity is defined as

$$\underset{H(X)}{Max} \left[H(X) - H(X \mid Y) \right]$$

where

$$H(X) = H(p1,p2,...) = -\sum_{1}^{I} p_i \, log \, (p_i)$$

and p_i, $i=1,2...I$, is the probability of transmitting x_i and $H(X \mid Y)$ is the condi-tional entropy of X given Y, i.e.,

$$H(X \mid Y) = \sum_{j=1}^{J} P[\, Y = y_j \,] H(X \mid Y = y_j).$$

By Shannon's theorem, if $H(X)$, the entropy of the source, is less than the channel capacity, then given any $\varepsilon > 0$ it is possible to transmit messages such that the probability of misclassifying a signal is less than ε. It follows that our goal in setting the channel's design parameters is to maximize its capacity or equivalently to minimize the loss of information. This loss can not be satisfactorily approximated by a quadratic function of the response i.e., the received signals. Because the actual values of X and Y, which can be non-numeric, have no significance at all and it is the probabilities (p_{ij}) that matter. So to take X and Y to be subsets of the real line and to base our analysis on these values will not result in conclusions with physical or engineering justifications. Of course, in the special case of binary channel by assigning values 0 and 1 to the transmitted and received signals we have

$$E[\,|\,f{-}g_0\,|_2] = \text{the probability of error}$$

where f is the response (received signal) and g_0 is the target (transmitted signal). Further, the binary symmetric channel

$$(p_{11} = p_{22} = p)$$

$$\text{Channel Capacity} = 1 - H(p, 1{-}p) = 1 + p\, logp + (1{-}p)\, log\,(1{-}p).$$

For this particular representation of the channel analysis of variance of the received and the transmitted signals has an F-ratio which is closely related to channel capacity nevertheless, this and the analysis based on it should not be viewed as a method for optimizing the design of a communication channel or a system with discrete input and output. In order to formulate the problem so that the method is applicable to it we need some knowledge of the inner working of the channel, such as the signals are voltages that are normally distributed etc., and in improving the design of the channel we should determine how various system parameters affect these distributions.

As was mentioned above, the two step optimization of Section 13.3 is based on the assumption that the loss from a system is measured by a quadratic function and the existence of a scaling factor; so the applicability of the methodology to a particular problem depends on the approximate validity of these assumptions. In practice, engineering knowledge and data analysis Nair[7] should be used to examine this, and if possible to reformulate the problem in such a way to make the method applicable to it.

13.6 CONCLUSION

We showed that in the case of quadratic loss function, the loss due to deviation of a system's response from its target can be decomposed into two components: a non-adjustable part which is related to the angular deviation of response from its desired target, and an adjustable one which is usually easy to correct. This loss can be reduced through proper setting of the system's parameters and the existence of an adjustable factor simplifies this operation; because in such cases we first maximize the S/N ratio and then eliminate any adjustable component of the loss. We also gave a geometric interpretation of this two step procedure, and by considering a communication channel illustrated the need for verifying the validity of certain assumptions before applying this procedure to an optimization problem.

APPENDIX

Here we show that the distance between the response and its target are the norms induced by $k(M) P_2(M)$ and $P_1(x)$ the measures on the signal and noise spaces.

Recall the P-diagram of Section 13.2. The ideal state of the system is when the response does not depend on x, the noise i.e.,

$$f(M,x;z) = g_0(M).$$

However, in practice the response deviates from its target, due to noise, and it is natural to ask how close, on the average, the system is to its intended target. In Section 13.2 we used a quadratic loss function and based our measurement on the expected total loss. In this case if $dP_2(M)$ and $dP_1(x)$ are the probability measures on the space of input and noise respectively then $d\mu = k(M) dP_2(M) dP_1(x)$ is a measure on the product space of the input and noise and the expected total loss is given by (x.1). Note that by taking the product measure we assume the noise and the input are independent. Further, it is a standard result that the following is a positive hermitian form on the space of squared integrable functions of signal and noise

$$<g_1 . g_2> = \int (g_1 - g_2)^2 \, d\mu.$$

Also $|| g || = <g . g>^{\frac{1}{2}}$ is a pseudo-norm and it is a standard result that this space can be transformed into an inner product space with the induced norm. And in this space the angle between two elements g_1 and g_2 is

$$Cos (g_1, g_2) = \frac{<g_i \cdot g_2>}{|| g_1 ||^{\frac{1}{2}} || g_2 ||^{\frac{1}{2}}} .$$

For further results see Krall (1973).[8]

REFERENCES

1. Phadke, M. S., and K. Dehnad. 1986. *Two Step Optimization for Robust Product and Process Design*. Preliminary Report.

2. Kackar, R. N. 1985. Off-Line Quality Control, Parameter Design, and the Taguchi Method. *Journal of Quality Technology* **17**:176-206.

   (Quality Control, Robust Design, and the Taguchi Method; Article 4)

3. Leon, Ramon V., Anne C. Shoemaker, and Raghu N. Kackar. 1987. "Performance Measures Independent of Adjustment: An Explanation and Extension of Taguchi's Signal-to-Noise Ratios." *Technometrics* vol. 29, no. 3 (Aug.):253-265.

   (Quality Control, Robust Design, and the Taguchi Method; Article 12)

4. Taguchi, Genichi, and M. S. Phadke. 1984. *Quality Engineering Through Design Optimization*.

   (Quality Control, Robust Design, and the Taguchi Method; Article 5)

5. Taguchi, G. and Y. Wu. 1980. *Introduction to Off-line Quality Control Systems*. Tokyo: Central Japan Quality Control Association.

6. Phadke, M. S., R. N. Kackar, D. V. Speeney, and M. J. Grieco. 1983. "Off-line Quality Control in Integrated Circuit Fabrication Using Experimental Design." *The Bell System Technical Journal* vol. 62, no. 5 (May-June):1273-1309.

   (Quality Control, Robust Design, and the Taguchi Method; Article 6)

7. Nair, V. N., and D. Pregibon. 1986. *A Data Analysis Strategy for Quality Engineering Experiments*. Preliminary Report.

   (Quality Control, Robust Design, and the Taguchi Method; Article 14)

8. Krall, A. M. 1973. *Linear Methods of Applied Analysis*. New York: Addison-Wesley.

14

A DATA ANALYSIS STRATEGY FOR
QUALITY ENGINEERING EXPERIMENTS

Vijay Nair
Daryl Pregibon

14.1 INTRODUCTION

Experimental design methods have traditionally focused on identifying the factors that affect the level of a production/manufacturing process. The Japanese, in particular Genichi Taguchi,[1] have demonstrated that for quality improvement, we also need to identify factors that affect the variability of a process. By setting these factors at their "optimal" levels, the product can be made robust to changes in operating and environmental conditions in the production line. Thus both the location and the dispersion effects of the design factors[1] are of interest. The present paper deals with techniques for analyzing data from quality engineering experiments for optimizing a process with fixed target.

We propose a structured data analytic approach with three phases of analysis: an exploratory phase, a modeling phase and an optimization phase. The focus in the exploratory phase is on determining the need for transforming the data. Graphical methods are used to determine the type of transformation, to assess location and dispersion effects, and to detect possible irregularities in the data. In the modeling phase, standard analysis of variance techniques, supplemented with probability plots of estimated effects, are used to identify the important design factors. In the optimization phase, the model is interpreted, the optimal

---

1. Glossary equivalent is *control parameters.*

levels of the factors are determined, and a feasible factor level combination is determined to optimize process quality.

Taguchi and Wu[1] recommend modeling the signal-to-noise ratio to determine the important dispersion effects. This is nearly equivalent to applying a logarithmic transformation to the data and modeling the variance of the transformed data. We propose a more general approach where data-analytic methods can be used to infer the appropriate transformation.

This paper is organized as follows. Section 14.2 provides an overview of our three-phase data analysis strategy. Section 14.3 describes an experiment conducted at AT&T Bell Laboratories to optimize the process of forming contact windows in complementary metal-oxide semiconductor circuits (Phadke and others[2]). Sections 14.4 through 14.6 discuss each of the three phases of the data analysis strategy in detail, where the data from the contact window example is used to demonstrate the techniques. In Section 14.7, some remarks are included on maximum likelihood techniques, the analysis of nonreplicated data, using other measures besides mean and variance, analysis of ordinal data, and software availability.

14.2 OUR STRATEGY: RATIONALE AND OVERVIEW

14.2.1 THE PARAMETER DESIGN PROBLEM

Consider the parameter design problem for optimizing a process with a fixed target (Taguchi and Wu;[1] Leon, Shoemaker, and Kackar[3]). We are interested in designing a process whose output Y is a function of two sets of factors: design factors — factors that can be controlled and manipulated by the process engineer, and noise factors[2] — all the uncontrollable factors including variability in the operating and environmental conditions. We use the notation $\mathbf{d}$ and $\mathbf{n}$ to denote the settings of the design and noise factors respectively. The target value for this process is fixed at t_0. If we can specify the cost when the output Y deviates from the target formally in terms of a loss function $L(Y, t_0)$, then the goal of the experiment is to determine the settings of the design factors $\mathbf{d}$ — parameter design — to minimize the average loss $E_{\mathbf{n}}\{L(Y, t_0)\}$ where $Y = f(\mathbf{d}, \mathbf{n})$.

Typically, the loss function cannot be specified precisely. Moreover, the form of the loss function could depend on the unknown transfer function $f(\cdot)$. Less formally then, the parameter design problem is to select the settings of the design factors to make the output Y as close as possible to the target t_0. In this paper, we interpret this as choosing $\mathbf{d}$ to minimize the variance of Y, $v_Y(\mathbf{d})$, subject to the

---

2. Glossary equivalent is *noise parameters*.

constraint that the mean of Y, $m_Y(\mathbf{d})$, is as close as possible to t_0. See the discussion in Section 14.7.2 for the use of other measures of location and dispersion.

14.2.2 THE ROLE OF DATA TRANSFORMATIONS

Before analyzing the data to determine the important design factors, we must first determine the extent to which the variance depends upon the mean. For if $v_Y(\mathbf{d}) = \gamma\,(m_Y(\mathbf{d}))$ for some function $\gamma\,(\cdot)$, then the variability of the process is completely determined by the constraint that the mean should be close to t_0. Nothing can be done to minimize the variability further.

Typically, the design factors $\mathbf{d}$ can be separated into two sets $\mathbf{d}_1$ and $\mathbf{d}_2$ so that the mean m_Y is a function of possibly both $\mathbf{d}_1$ and $\mathbf{d}_2$ while the variance v_Y can be factored as

$$v_Y(\mathbf{d}) = \gamma\,(m_Y(\mathbf{d}_1, \mathbf{d}_2))\sigma^2(\mathbf{d}_2). \tag{1}$$

To determine the effects of the design factors on variability, we should restrict our attention to the component $\sigma^2(\mathbf{d}_2)$ in equation (1). The remaining contribution to the variability is determined by the constraint that the mean has to be on target. In fact, if one or more of the factors $\mathbf{d}_1$ (called adjustment factors,[3] Leon, Shoemaker, and Kackar[3]) are known a priori, we need to model only the factors $\mathbf{d}_2$ to minimize $\sigma^2(\mathbf{d}_2)$. The mean can then be brought on target by manipulating the adjustment factors.

Suppose the function $\gamma\,(\cdot)$ is known. Then one approach to analyzing the dispersion effects is to estimate $\sigma^2 = v_Y/\gamma\,(m_Y)$ and model it as a function of the design parameters. Taguchi's (Taguchi and Wu[1]) signal-to-noise ratio analysis is a special case of this with $\gamma(m) = m^2$. There is no reason, however, that this special case should always hold.

An alternative approach for analyzing the dispersion effects in equation (1), and that which we recommend, is to transform Y to $Z = \tau(Y)$ where $\tau'(Y) = 1/[\gamma\,(Y)]^{\frac{1}{2}}$. Then it can be shown by a Taylor series argument that $v_Z \approx \sigma^2$. We can then estimate the variance of Z and model it as a function of the design factors. Such transformations are known in the statistical literature as variance-stabilizing transformations (Bartlett[4]). Empirical evidence also suggests that often these transformations have other side benefits such as enhancing both the symmetry of the underlying (noise) distribution and the additivity of the mean as a function of the design factors. For this reason, when $\gamma(m) = m^2$ so that Taguchi's signal-to-noise analysis is appropriate, we recommend that a logarithmic

---

3. Glossary equivalent is *adjustment parameters*.

transformation be applied to the data. The mean and the variance of the transformed data can then be analyzed separately to determine the location and dispersion effects.

The real difficulty lies in diagnosing the form of the unknown function $\gamma(\cdot)$ in equation (1) from the data. When there are a sufficient number of replications so that both the mean and variance of Y can be estimated with a reasonable degree of accuracy, it is possible to use data-analytic methods to determine the form of $\gamma(\cdot)$.

14.2.3 OVERVIEW OF STRATEGY

Figure 14.1 gives a schematic representation of our recommended analysis strategy. We begin the exploratory phase with a plot of the data to check for possible irregularities, and to visually assess the differences in the means and variances from the different factor-level settings. To determine the need for a variance-stabilizing transformation, we plot the variances versus the means. If no strong functional relationship is apparent, we do not transform the data and proceed to the next phase to model the dispersion and location effects separately. If there is a strong relationship, we use the plot to guide the choice of transformation, and repeat the process on the transformed data.

In the modeling phase, we recommend supplementing the standard analysis of variance computations by decomposing the effects into meaningful single degree-of-freedom contrasts. Probability plots can then be used to determine the important design factors. If it is important to obtain good estimates of the effects, a parametric model can be fitted to the data by maximum likelihood methods.

In the optimization phase, the optimal levels of the factors that affect the dispersion and those that affect the location are determined. From this a feasible factor-level combination that optimizes process quality is obtained. When there are adjustment factors (Taguchi and Wu[1]; Leon, Shoemaker, and Kackar[3]) that are known a priori, one needs to model only the dispersion effects. The mean can be made close to the target value by fine-tuning the adjustment factors. This two-step approach to optimization can also be used when the adjustment factors are not known a priori. Typically, during the modeling phase, we would discover some design factors that affect the location but not the dispersion. Given the "discovered" adjustment factors, we can first choose the levels of other factors to minimize the dispersion and then use the signal factors to bring the mean close to target.

14.3 AN EXAMPLE

The following experiment was conducted at AT&T Bell Laboratories to optimize the process of forming contact windows in complementary metal-oxide

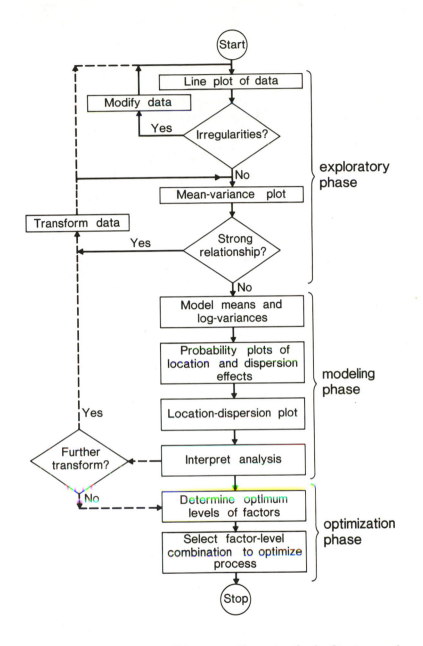

Figure 14.1. Flow Diagram Describing our Data Analysis Strategy. Arrowheads indicate the direction of flow. Diamonds represent decision points. Dotted lines indicate optional paths recommended for a thorough analysis.

semiconductor circuits (Phadke and others[2]). The contact windows facilitate inter-connections between the gates, sources, and drains in a circuit. The process of forming contact windows involves photolithography. Phadke and others,[2] identified nine factors that were important in controlling window sizes: A – mask dimension, B – viscosity, C – spin speed, D – bake temperature, E – bake time, F – aperture, G – exposure time, H – developing time, and I – plasma etch time. Two levels were chosen for factors A, B, and D, and all others were at three levels. The authors decided to combine factors B and D to form a three-level factor BD and use the L_{18} orthogonal array experimental design displayed in Table 14.1.

The target size for contact windows was in the range 3.0 to 3.5 μm. It was important to produce windows near the target dimension; windows not open or too small result in loss of contact to the devices while excessively large windows lead to shorted device features. The actual size of the windows in the functional circuits on a chip could not be accurately measured. Instead, there were three sur-rogate measures of quality: pre-etch and post-etch line width and post-etch win-dow size of test patterns provided in the upper left hand corner of each chip. Ten measurements were made at each experimental run: two wafers with five chips each corresponding to specific locations on a wafer — top, bottom, left, right and center. In this paper, we consider only the analysis of the pre-etch line width data given in Table 14.2. For this measure of quality, factor I, plasma etch time, is not a relevant design factor.

TABLE 14.1. EXPERIMENTAL DESIGN

Exp #	A	BD	C	E	F	G	H	I
1	1	1	1	1	1	1	1	1
2	1	1	2	2	2	2	2	2
3	1	1	3	3	3	3	3	3
4	1	2	1	1	2	2	3	3
5	1	2	2	2	3	3	1	1
6	1	2	3	3	1	1	2	2
7	1	3	1	2	1	3	2	3
8	1	3	2	3	2	1	3	1
9	1	3	3	1	3	2	1	2
10	2	1	1	3	3	2	2	1
11	2	1	2	1	1	3	3	2
12	2	1	3	2	2	1	1	3
13	2	2	1	2	3	1	3	2
14	2	2	2	3	1	2	1	3
15	2	2	3	1	2	3	2	1
16	2	3	1	3	2	3	1	2
17	2	3	2	1	3	1	2	3
18	2	3	3	2	1	2	3	1

TABLE 14.2. PRE-ETCH LINE WIDTH DATA

1	2.43	2.52	2.63	2.52	2.50	2.36	2.50	2.62	2.43	2.49
2	2.76	2.66	2.74	2.60	2.53	2.66	2.73	2.95	2.57	2.64
3	2.82	2.71	2.78	2.55	2.36	2.76	2.67	2.90	2.62	2.43
4	2.02	2.06	2.21	1.98	2.13	1.85	1.66	2.07	1.81	1.83
5	1.87	1.78	2.07	1.80	1.83					
6	2.51	2.56	2.55	2.45	2.53	2.68	2.60	2.85	2.55	2.56
7	1.99	1.99	2.11	1.99	2.00	1.96	2.20	2.04	2.01	2.03
8	3.15	3.44	3.67	3.09	3.06	3.27	3.29	3.49	3.02	3.19
9	3.00	2.91	3.07	2.66	2.74	2.73	2.79	3.00	2.69	2.70
10	2.69	2.50	2.51	2.46	2.40	2.75	2.73	2.75	2.78	3.03
11	3.20	3.19	3.32	3.20	3.15	3.07	3.14	3.14	3.13	3.12
12	3.21	3.32	3.33	3.23	3.10	3.48	3.44	3.49	3.25	3.38
13	2.60	2.56	2.62	2.55	2.56	2.53	2.49	2.79	2.50	2.56
14	2.18	2.20	2.45	2.22	2.32	2.33	2.20	2.41	2.37	2.38
15	2.45	2.50	2.51	2.43	2.43					
16	2.67	2.53	2.72	2.70	2.60	2.76	2.67	2.73	2.69	2.60
17	3.31	3.30	3.44	3.12	3.14	3.12	2.97	3.18	3.03	2.95
18	3.46	3.49	3.50	3.45	3.57					

14.4 PHASE I — EXPLORATORY ANALYSIS

It is important to begin the analysis with a visual display of the experimental data that summarizes the main features such as location and dispersion, and also highlights possible irregularities. The *line plot* in Figure 14.2 is a variation of the box plot (Tukey[5]) commonly used to display such information. For each of the experiments, the line plot displays the individual mean and standard deviation, and all replicates which fall outside $\pm\sigma$ interval about the mean.

The line plot is useful for identifying individual replicates within an experiment which are several standard deviations away from the norm. These could possibly be "bad" data points, in which case they should be omitted from further analysis. For the pre-etch line width data, there are no apparent irregularities in the data. If some observations are discarded, we recommend that the line plot be redone. This ensures that apparent differences in location and dispersion are due to real effects and not "bad" data.

The line plot can also be used to visually assess differences in location and dispersion across the different experiments. Figure 14.2 shows that the between-experiment variability is large compared to the within-experiment variability. Thus we can expect to find large location effects in the modeling phase. The $\pm\sigma$ intervals appear to vary across experiments, but not dramatically so. This suggests the presence of only moderate dispersion effects.

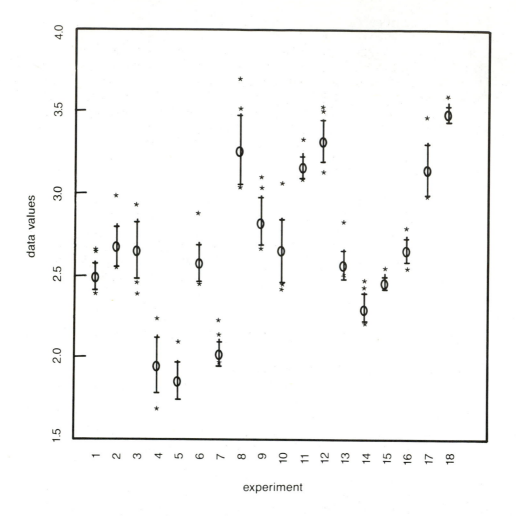

Figure 14.2. Line Plot of the Pre-Etch Data. For each of the eighteen individual experiments, the symbol "O" denotes the sample mean, the line extends ± 1 s.d. about the mean, and the symbol "*" denotes individual values which lie outside this interval. There are obvious location effects in the data, and possibly some dispersion effects.

The next step is the *mean-variance plot* shown in Figure 14.3. The means and the variances from the different runs are plotted on a log-log scale. If $\sigma^2(\mathbf{d}_2)$ in equation (1) is constant, i.e. it does not depend on the design factors, the mean-variance plot will reveal the form of γ ($\cdot$). If $m_Y(\mathbf{d})$ is constant, the plot will show no relationship between mean and variance. The log-log scale is used since the plot will be approximately linear with slope k for the common situation where γ $(m)=m^k$.

Typically, however, both σ and m_Y will depend on the design factors to some extent. It may not be possible, in general, to separate γ ($\cdot$) and $\sigma(\cdot)$ in equation (1). But if γ $(m_Y(\mathbf{d}))$ is the dominant component, the mean-variance plot will still be useful in detecting the relationship.

If the plot suggests a strong relationship, we can fit a line through the plot (say by eye) and estimate k from the slope of the line. If $k \approx 0$, there is no strong relationship and no transformation is necessary. Values of k near 1, 2, and 4 correspond respectively to the square-root, logarithmic, and reciprocal transformations. These are the three common transformations found to be useful in practice. From Figure 14.3, we see that there is no strong relationship and so no transformation is necessary. We can now proceed to the next phase and model the means and variances. If we had found the need for transformation, we should repeat the above cycle with the transformed data (see Figure 14.1).

14.5 PHASE II — MODELING

14.5.1 DISPERSION EFFECTS

To determine the important dispersion effects, we model the logarithms of the variances of the (possibly transformed) data as an additive function of the design factors. Additivity is more likely to be satisfied on the logarithmic scale, and moreover, allows estimation of dispersion effects by unconstrained least-squares.

Standard analysis of variance techniques (Box, Hunter, and Hunter[6]) and the resulting F-tests can be used to determine the significant factors when an estimate of error is available. Otherwise, one must resort to less formal probability plotting techniques. However, we recommend that the probability plotting methods be used to supplement the formal ANOVA computations and the F-tests in all cases. If we perform many F-tests, as would be the case in screening experiments with many factors, we would find some of these to be significant even when none of the factors have any effects. The probability plotting method compensates for this appropriately.

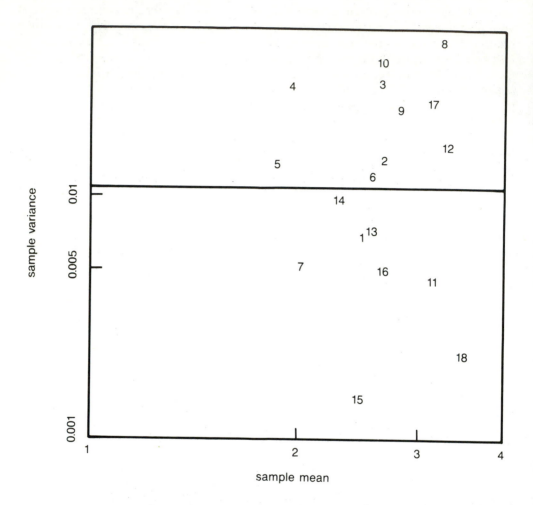

Figure 14.3. Mean-Variance Plot of the Pre-Etch Data. On a log-log scale, we plot the sample variance from each experiment versus the corresponding sample mean. The experiment number is used as the plotting character. The line superimposed on the scatter plot was fitted by eye. There appears to be no significant association between variance and mean, linear or otherwise.

To do a probability plot, we first decompose factors at more than two levels into meaningful single degree-of-freedom contrasts. For quantitative factors, this would typically be the components that are linear, quadratic, and so forth. The half-normal probability plot (Daniel[7]) is then obtained by plotting the ordered absolute values of the contrasts against the half-normal quantiles. A linear configuration through the origin suggests that there are no significant contrasts. If there are a few important effects, they will tend to fall towards the top right-hand corner above the linear configuration determined by the rest of the contrasts. Thus the probability plot uses the variability among the different contrasts to detect the few important ones.

The results from the ANOVA computations for the pre-etch line width data are given in Table 14.3. The F-tests show that none of the factors is really significant. Factors A, F, and G have the largest observed effects. Figure 14.4 is the half-normal probability plot of the single degree-of-freedom dispersion contrasts. The linear component of BD measures the effect of D. The quadratic component measures the effect of B provided D has no effect. Since this appears to be the case, we have denoted these terms as B and D in Figure 14.4. For the other three-level factors, the components are the usual linear and quadratic terms. The overall linear appearance of the plot suggests that there are no strong dispersion effects. But the effects of factors F, A and G are somewhat separated from the others. Notice also that by decomposing the factors into linear and quadratic terms, we have not diffused the linear effects of F and G.

Although the dispersion effects of F, A, and G appear to be only marginally important, we will include these factors in the optimization phase. The possible error involved in making this decision is less costly than in wrongly concluding that the factors have no effect.

TABLE 14.3. ANOVA TABLE FOR DISPERSION EFFECTS

	df	SS	MS	F
A	1	0.493	0.493	4.059
BD	2	0.121	0.060	0.498
C	2	0.150	0.075	0.615
E	2	0.380	0.190	1.563
F	2	0.831	0.415	3.421
G	2	0.496	0.248	2.042
H	2	0.011	0.006	0.045
Residual	4	0.486	0.121	

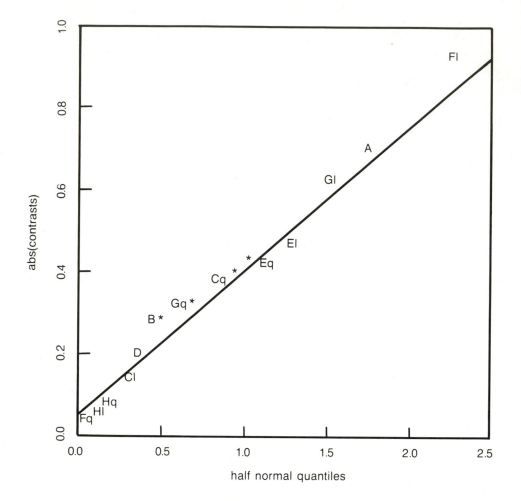

Figure 14.4. Probability Plot of Dispersion Effects for the Pre-Etch Data. The absolute values of the single degree-of-freedom contrasts are plotted against the quantities of the half-normal distribution. The factor level combination is used as the plotting character. The contrasts labeled "*" correspond to "error contrasts" rather than effects of interest. The line superimposed on the plot was fitted by eye. The plot shows that dispersion effects *Fl*, *A*, and *Gl* are marginally important.

14.5.2 LOCATION EFFECTS

The important location effects can be determined by modeling the means of the (possibly transformed) data as an additive function of the design factors. It is possible to do a careful and efficient weighted a least-squares analysis that takes into account the results from the previous section on the differences in variances. For the sake of simplicity, we will restrict attention to a least-squares analysis which would be adequate in detecting the really important effects. See, however, the "Maximum Likelihood Analysis" section for a more efficient analysis. The F-tests and the probability plots in this section should also be considered as approximate procedures.

The ANOVA table for the location effects for the pre-etch line width data are given in Table 14.4. In addition to the residual sum of squares, we have provided the within replications sum of squares which also provides an "error" estimate. We use the term "error" here as some average measure of the underlying variability since the variances are possibly different. The within replications mean square in Table 14.4 is smaller than the residual mean square error. To be conservative, the F-statistics in Table 14.4 are computed with the residual mean square error. They suggest that the location effects associated with factors A, BD, and, to a lesser extent, C and G are significant. The next largest observed effect is due to factor H.

TABLE 14.4. ANOVA TABLE FOR LOCATION EFFECTS

	df	SS	MS	F
A	1	0.651	0.651	22.459**
BD	2	1.345	0.672	23.186**
C	2	0.765	0.383	13.193*
E	2	0.002	0.001	0.038
F	2	0.032	0.016	0.545
G	2	0.545	0.273	9.397*
H	2	0.001	0.110	1.030
Residual	4	0.116	0.029	
Within reps	147	2.481	0.017	

** $p < 0.01$

* $p < 0.05$

Figure 14.5 shows the half-normal probability plot of the single degree-of-freedom contrasts. We see that A, B, and the linear components of C, G, and H are the important location effects.

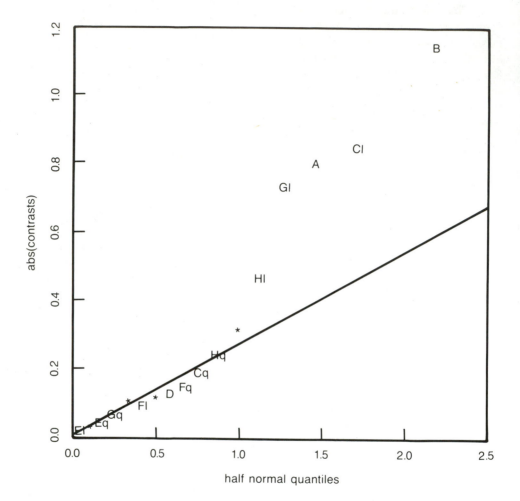

Figure 14.5. Probability Plot of Location Effects of the Pre-Etch Data. The absolute values of the single degree-of-freedom contrasts are plotted against the quantities of the half-normal distribution. The factor level combination is used as the plotting character. The contrasts labeled "*" correspond to "error contrasts" rather than effects of interest. The line superimposed on the plot was fitted (to the nonsignificant contrasts) by eye. The plot shows that location effects *B*, *Cl*, *A*, *Gl*, and to a lesser extent *Hl*, are important.

14.5.3 INTERPRETING THE ANALYSIS

The results from the modeling phase can be summarized in the *location-dispersion plot* shown in Figure 14.6. The single degree-of-freedom dispersion contrasts are plotted against the corresponding location contrasts, and the shaded regions, obtained from the probability plots, indicate effects that are not important. Thus the factors in the intersected area of Figure 14.6 exhibit neither location nor dispersion effects; those lying in the horizontal (vertical) band exhibit location (dispersion) effects only. Those in the unshaded areas have both location and dispersion effects. For the pre-etch data, B, C, and to a lesser extent H are the adjustment factors, i.e. factors with only location effects. F is a pure dispersion factor while A and G have both location and dispersion effects. Further, only the linear terms of the three-level factors are important; the quadratic terms appear to be insignificant.

The location-dispersion plot can also be used to assess the effectiveness of the data transformations in the exploratory phase. If most of the important factors have both location and dispersion effects and they fall in either the top right-hand or the lower left-hand unshaded areas in Figure 14.6, we should suspect that the variance of our (possibly transformed) data is an increasing function of the mean. Thus a further transformation may be necessary (see Figure 14.1). Even if none of the factors is important, we should examine the location-dispersion plot to see if the location and dispersion estimates appear to be positively (negatively) correlated. Strong positive (negative) correlation suggests that the underlying noise distribution is very skewed to the right (left). It may then be worthwhile to transform the data and see if there are any important effects in the transformed scale. For the pre-etch data, Figure 14.6 does not indicate the need for a transformation.

14.6 PHASE III — OPTIMIZATION

We can use a two-step approach to optimization if there are adjustment factors, i.e. factors with only location effects, that are either known a priori or are discovered during the modeling phase. In the first step, we can set the factors with dispersion effects at the optimal levels to minimize variability. The mean can then be brought on target by fine-tuning the adjustment factors.

For the pre-etch line width data, as seen in the preceding section, B, viscosity, and C, spin speed, are the adjustment factors while F, aperture, is a pure dispersion factor. A, mask dimension, and G, exposure time, have both location and dispersion effects. Figure 14.7, the *factor effects plot*, provides both the magnitude and the direction of these effects. This information can also be gleaned from the location-dispersion plot if the factors have only linear effects. We see from Figure 14.7 that the optimal levels for minimizing dispersion are: 1 for F, 2 for A, and 3 for G. At these optimal levels, the location effects of A and G effectively cancel each other. For the adjustment factors, the location decreases with viscosity, B, while it increases with spin speed, C.

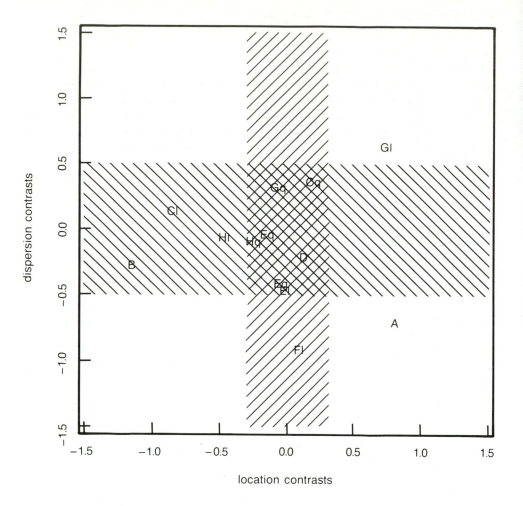

Figure 14.6. Location-Dispersion Plot of Effects for the Pre-Etch Data. The single degree-of-freedom contrasts for dispersion are plotted against those for location. The factor-level combination is used as the plotting character. The contrasts labeled "*" correspond to "error contrasts" rather than effects of interest. The shaded bands are derived from the individual probability plots of location and dispersion effects. Contrasts appearing in the cross-hatched portion of the plot are those which exhibit neither location nor dispersion effects. Those lying in the horizontal (vertical) band exhibit location (dispersion) effects only. Those contrasts appearing in unshaded areas exhibit both location and dispersion effects.

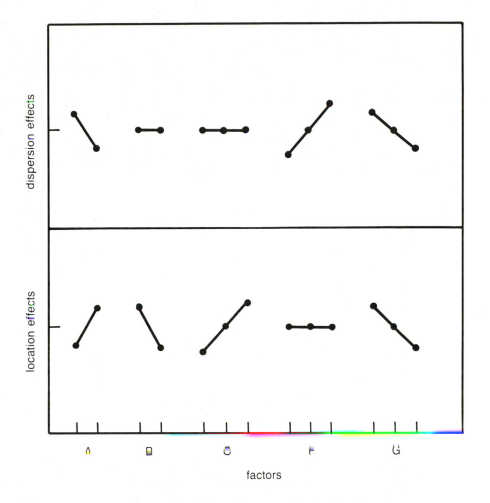

Figure 14.7. Factor-Level Effects Plot for the Pre-Etch Data. For each of the important factors identified by our suggested analysis, the estimated effect for each level of the factor is displayed for both location and dispersion. The plot is useful for determining the optimal levels of the factors. Note that the effects are all linear since none of the quadratic components of the three-level factors were important.

We can now choose between F, A, or G (or even a combination of these) to minimize dispersion. Note also that G is a quantitative variable so that we may be able to reduce variability even further by linearly extrapolating beyond level 3. However, the final choice of the factor levels should be guided by engineering as well as statistical considerations.

The standard operating levels for the factors A, B, C, F, and G were, respectively, 1, 1, 2, 2, and 2. Purely on the basis of our statistical analysis, we would have recommended that the level of A be changed to 2, C be changed to 3, F be changed to 1 and, possibly, G be changed to 3. Phadke and others[2] decided not to change the level of F because of engineering considerations. In addition, they did not recommend changing the level of C. Their conclusions were based on a signal-to-noise ratio analysis of not only the data we consider here (pre-etch line width), but also the other two measures of quality (post-etch line width and window size).

In cases where there are no adjustment factors or where the adjustment factors cannot be easily manipulated, the optimization process is more complicated. The optimal levels of the dispersion factors and the location-dispersion factors have to be determined simultaneously. It is possible to formulate this formally as a mathematical problem and use a constrained optimization technique to solve it. However, a simple iterative search procedure over the desirable factor space is likely to be more useful.

14.7 DISCUSSION

14.7.1 MAXIMUM LIKELIHOOD ANALYSIS

The analyses discussed in "Phase II: Modeling" are geared primarily to the identification of important effects, both for location and dispersion. As a byproduct of the computations used in these analyses, estimates of the effects are available. By appropriate choice of data transformation, these estimates should be good, though not generally optimal. The problem derives primarily from the fact that the optimal estimates of location effects depend upon the presence or absence of dispersion effects. In particular, the unweighted, orthogonal array analysis of sample means we carried out is formally incorrect, since we are weighting each sample mean equally, though the presence of dispersion effects dictates otherwise.

A more formal method of estimation can be carried out (Nelder and Pregibon[8]). What results is an iterative method of estimation, alternating between estimating location effects by weighted least-squares, with weights depending on the current estimates of dispersion effects, and estimating dispersion effects by least-squares (as suggested), where now dispersion is measured about current location effects rather than sample means. This more detailed analysis requires using all the data, not just the sample means and variances of the individual experiments. We don't recommend this procedure in general, but if the results of the analysis are to be used for producing forecasts and balance sheets, then the extra computational effort is strongly recommended.

14.7.2 ANALYSIS OF NONREPLICATED DATA

The data analytic approach developed in this paper assumes that there is a sufficient number of replications at each experimental setting. Some of our suggestions, particularly those in the exploratory phase, either have to be modified or cannot be used when there are no replications. Little if anything can be inferred from the data about transformations. The appropriate scale of analysis must be determined by the experimenter based on similar previous analyses and/or insight into the physical mechanism giving rise to the observed data. To estimate dispersion effects, one has to first identify the important location effects. Assuming all other location effects to be zero induces pseudo-replication in the data, allowing the methodology discussed in the present paper to be applied. For more discussion on this topic, see Box and Meyer[9] and Nair and Pregibon[10].

14.7.3 OTHER MEASURES OF LOCATION AND DISPERSION

We have limited our discussion to the mean and variance as measures of location and dispersion respectively, mainly due to their familiarity among engineers. It should be noted however that similar analyses can be carried out with other measures, and it may even be desirable to do so. For example, it may be more meaningful to choose the design factors so that the *median*, rather than the *mean*, is close to the target t_0. Similarly, the *interquartile range*, the range of the central 50% of the data, may describe dispersion better than *variance*. Apart from such subject matter considerations, an important property of measures such as the median and interquartile range is that their sample-based estimates are more resistant to "bad" data points than those corresponding to the mean and variance.

14.7.4 ANALYSIS OF ORDERED CATEGORICAL DATA

In some cases, the response variable in a quality improvement study consists of categorical data with an implied ordering in the categories. Taguchi has proposed a method called "accumulation analysis" for analyzing such data. See Nair[11] and the discussion following that paper for the properties of this method and for more methods of analysis. McCullagh[12] discusses fitting parametric models to ordered categorical data by maximum likelihood techniques.

14.7.5 SOFTWARE

The analysis reported in this paper was carried out entirely in the S system and language for data analysis (Becker and Chambers[13]). S is an environment for statistical computing and graphics, but as such, has no special facilities for the analysis of designed experiments. A special purpose *macro* package (Chambers and Freeny[14]) provided the calculations we required.

The flow-diagram in Figure 14.1 explicitly describes how we feel an analysis should proceed. This expertise can be coded into software, forming the "knowledge-base" of so-called *expert systems*. We have limited experience with this type of software, though a prototype system for linear regression analysis, REX, has been developed (Gale and Pregibon[15]). Of particular relevance is the fact the knowledge coded into REX is exactly of the sort displayed in Figure 14.1 and described in more detail in previous sections. We thus foresee the possibility of providing an intelligent interface to S to help engineers model their data and optimize process quality.

14.8 CONCLUSION

In quality engineering experiments, there is typically little interest in detailed structural modeling and analysis of the data. The user's primary goal is to identify the really important factors and determine the levels to optimize process quality. For these reasons, we have tried to keep our recommended strategy fairly simple. It can be supplemented with more sophisticated techniques depending on the needs and statistical training of the user. The availability of software would also reduce the need for simplicity. Finally, we note that we have not considered the analysis of data from the confirmation experiment in our data analysis strategy since this analysis is routine. It should be emphasized, however, that the confirmation experiment to determine if the new factor levels do improve process quality is an integral part of quality engineering experiments.

REFERENCES

1. Taguchi, G., and Y. Wu. 1980. *Introduction to Off-line Quality Control.* Japan: Central Japan Quality Control Association.

2. Phadke, M. S., R. N. Kackar, D. V. Speeney, and M. J. Grieco. 1983. "Off-line Quality Control is Integrated Circuit Fabrication Using Design of Experiments." *The Bell System Technical Journal* vol. 62, no. 5 (May-June):1273-1309.
   (Quality Control, Robust Design, and the Taguchi Method; Article 6)

3. Leon, Ramon V., Anne C. Shoemaker, and Raghu N. Kackar. 1987. "Performance Measures Independent of Adjustment: An Explanation and Extension of Taguchi's Signal-to-Noise Ratios." *Technometrics* vol. 29, no. 3 (Aug.):253-265.
   (Quality Control, Robust Design, and the Taguchi Method; Article 12)

4. Bartlett, M. S. 1947. "The Use of Transformations." *Biometrics* **3**:39-52.

5. Tukey, J. W. 1977. *Exploratory Data Analysis*. Addison Wesley.

6. Box, G. E. P., W. G. Hunter, and J. S. Hunter. 1978. *Statistics for Experimenters*. New York: John Wiley and Sons.

7. Daniel, C. 1959. "Use of Half-Normal Plots in Interpreting Factorial Two-Level Experiments." *Technometrics* vol. 1, no. 4 (Nov.).

8. Nelder, J. A., and D. Pregibon. n.d. "Quasi-likelihood Models." *Biometrika*. Forthcoming.

9. Box, G. E. P., and D. Meyer. 1985. "Studies in Quality Improvement I: Dispersion Effects from Fractional Designs." *MRC Technical Report #2796* (Feb.). Madison, Wis.: University of Wisconsin.

10. Nair, V. N., and D. Pregibon. 1985. "Analysis of Dispersion Effects: When to Log?" *AT&T Bell Laboratories Technical Memorandum* (Dec.).

11. Nair, V. N. 1985. "Testing in Industrial Experiments with Ordered Categorical Data." *AT&T Bell Laboratories Technical Memorandum* (June).
(Quality Control, Robust Design, and the Taguchi Method; Article 11)

12. McCullagh, P. 1980. "Regression Models for Ordinal Data (with discussion)." *Journal of the Royal Statistical Society, Ser. B* **42**:109-142.

13. Becker, R. A., and J. M. Chambers. 1984. *S: An Interactive Environment for Data Analysis and Graphics*. Belmont, Calif.: Wadsworth.

14. Chambers, J. M., and A. E. Freeny. 1985. "A Decentralized Approach to Analysis of Variance." *AT&T Bell Laboratories Technical Memorandum* (Sept.).

15. Gale, W. A., and P. Pregibon. 1984. "REX: An Expert System for Regression Analysis." *Proceedings of COMPSTAT 84*, 242-248. Prague (Sept.). Pub. Phsyica-Verlag.